ADVANCED COOLING TECHNOLOGIES AND APPLICATIONS

Edited by **S. M. Sohel Murshed**

Advanced Cooling Technologies and Applications
http://dx.doi.org/10.5772/intechopen.74636
Edited by S. M. Sohel Murshed

Contributors

Ri Li, Xuan Gao, Juergen Suess, Florian Hanslik, Vinicius Silva, Miguel Edgar Morales Udaeta, S. M. Sohel Murshed, Bin Chen, Zhifu Zhou, Liwen Jin, Xin Cui, Xiaohu Yang, Yanjun Sun, Xiangzhao Meng, Fangming Jiang, Jiwen Cen, Zhibin Li, Yiwei Wang, Shaoxiong Liao, Fuwen Liang, Ashok Dewangan, Anil Kumar, Ravi Kumar

First published in London, United Kingdom, 2019 by IntechOpen
IntechOpen is the global imprint of INTECHOPEN LIMITED, registered in England and Wales, registration number: 11086078, The Shard, 25th floor, 32 London Bridge Street
London, SE19SG – United Kingdom
Printed in Croatia

British Library Cataloguing-in-Publication Data
A catalogue record for this book is available from the British Library

Additional hard copies can be obtained from orders@intechopen.com

Advanced Cooling Technologies and Applications, Edited by S. M. Sohel Murshed
p. cm.
Print ISBN 978-1-78984-838-0
Online ISBN 978-1-78984-839-7

For EU product safety concerns:
IN TECH d.o.o., Prolaz Marije Krucifikse Kozulić 3, 51000 Rijeka, Croatia,
info@intechopen.com or visit our website at intechopen.com.

Meet the editor

Professor S. M. Sohel Murshed was born in Bangladesh and obtained his PhD in Mechanical and Aerospace Engineering from Nanyang Technological University, Singapore, in 2007. He is currently a professor of thermofluids and energy conversion technologies at the Department of Mechanical Engineering of IST, University of Lisbon, Portugal. Previously he worked as a postdoctoral fellow at universities in Singapore and the USA. He has so far authored/coauthored seven books, 21 book chapters, and over 125 papers in leading international journals and conferences. Several of his papers have been classified as Highly Cited Papers by Web of Science and his current Google Scholar citation count is 4350 with an h-index of 25. He is a Portugal national delegate to the management committee of a European COST Action on nanofluids and is also a group leader of the Action. He is an editor of the *Journal of Nanofluids* (ASP) and an associate editor of the *Journal of Thermal Analysis and Calorimetry* (Springer). He holds membership of various organizations such as ASME and ASTFE. His main research interests include nanofluids, micro- and nanoscale heat transfer, heat exchangers, microfluidics, advanced energy, and cooling technologies.

Contents

Preface

In recent years, research and development in cooling technologies has attracted great attention worldwide and good progress has been made as numbers of new cooling systems have emerged. However, the fast advances in many high-tech products and high-performance devices together with the current trends of miniaturization have resulted in very high heat generation of those products and devices leading to serious technical challenges for their fast and adequate cooling, whereas conventional cooling techniques are falling short in dealing with such high thermal management. Thus, advanced and innovative cooling technologies are of immense importance to meet the ever-increasing cooling demand of those modern devices, systems, and processes. This book covers research and development of several major areas of advanced cooling approaches, their performance, challenges, as well as their applications, and consists of eight chapters.

The book starts with a short chapter providing a brief overview of advanced cooling technologies and also highlighting all the contributions. The second chapter presents a novel energy-saving and environmentally-friendly air-conditioning method, which is an indirect evaporative air cooling based on the M-cycle. It also reports both experimental and computational works and findings of this heat exchange system. Spray cooling is one of the effective cooling technologies for high heat flux removal requirements and this cooling technique is analyzed from two aspects: the entire spray (spray level) and droplets (droplet level) both of which are thoroughly reviewed in the third chapter. The fourth chapter introduces development and evaluation of a heat pump-based innovative energy system particularly for cooling high-power LED lamps and also for waste heat recovery. The findings of this study reveal that this novel cooling system can lead to significant energy savings. The fifth chapter reports design and development of a modular architecture-based cooling system for photovoltaic plant. A detailed analysis of the regional climatic, geographic, and solar conditions, as well as construction, operational, and maintenance aspects, is also provided in this chapter. Boiling is one of the most effective cooling mechanisms and an experimental study on nucleate pool boiling performance of refrigerants on a powder-coated heating surface is presented in Chapter 6. Besides evaluating performance, it also presents an empirical correlation for prediction of the boiling heat transfer coefficient of refrigerants. The last two chapters focus on specific and industrial applications of cooling technologies. While Chapter 7 reports fundamental applications of cryogen spray cooling (transient flashing) in laser dermatology, water-based centrifugal compressor cooling systems for industrial applications is carefully studied in the final chapter.

The book is intended to provide useful information and be a guide to students, researchers, academicians, engineers, manufacturers, and other professionals who are involved or interested in the areas of cooling technologies, thermal management systems, and their applications.

I would like to especially thank all the authors for their contributions and the staff from the publisher for providing continuous support, which have made possible the successful completion of this book.

This book is dedicated to my parents and family, especially to my two-year-old daughter (Samara) who has been the source of our joy and happiness.

S. M. Sohel Murshed
Instituto Superior Técnico
Universidade de Lisboa
Lisbon, Portugal

Chapter 1

Introductory Chapter: A Brief Note on Advanced Cooling Technologies

S M Sohel Murshed

Additional information is available at the end of the chapter

http://dx.doi.org/10.5772/intechopen.82340

1. Introduction

The smaller and faster based modern trend of manufacturing devices and equipment leads to dramatic increase in heat generation resulting in their higher failure rate and shorter longevity. For instance, the reduction of size of high-tech electronic devices and huge increase in the number of integrated components and subsequent increase in power density yielded enormous challenges for their fast and adequate cooling [1]. Since conventional cooling techniques are falling short in dealing with such high cooling demand, advanced and innovative cooling technologies are needed to meet the raising cooling demand of those modern devices, systems, and processes. In recent years, research and development in advanced cooling technologies as well as search for superior coolants have attracted tremendous interest worldwide and good progress has been made as numbers of new cooling systems as well as few new coolants have emerged [2, 3]. Recently, a feature on cooling technologies was published in technology report of BBC News where cooling technologies were also identified to be a red-hot sector [4]. While there is an urgent need to reduce (cool down) the global warming, the increasing cooling needs from modern devices, systems, and appliances as well as human comfort (like district heating and cooling) to be met with the advanced cooling technologies and coolants.

This chapter aims to provide a short note on advanced cooling technologies and emerging coolants as well as summarizing the key features and findings from each contributed chapter of this book.

2. Cooling technologies and emerging coolants

2.1. Cooling technologies

In order to employ cooling technologies for cooling any devices, components, equipment, process, appliance, or even space (rooms, buildings, etc.) for human comfort, it is necessary to understand the types of cooling technologies and their working mechanisms, capability, limitations, and other key information as well. Since most of these technologies are well established and explained in the literature and text books (e.g., [5, 6]), they will not be elaborated further here. However, based on energy input, a classification schematic of main cooling technologies and their subclasses is provided in **Figure 1**. It is noted that the cooling technologies can also be classified based on other factors such as coolants, and devices or systems to be cooled as well as direct or indirect cooling. For instance, depending on thermal load (heat flux), cooling methods, particularly for electronic devices and equipment, can generally be classified as follows:

- natural convection with air;
- forced convection with air;
- forced circulation of water;
- immersion cooling with natural convection;
- immersion cooling with boiling;
- heat pipe; and
- mini- or microchannel heat sinks (with liquid).

The selection of a cooling technique mostly depends on the heat generation by the equipment and the maximum heat transfer capacity of the method as well as their suitability. In general, natural convection is used where the heat dissipation is low, whereas for very high heat loads like in super computers and high heat-generating communication equipment, immersion cooling needs to be employed. For example, the generated heat flux of electronic components used in high-power microwave equipment such as radars can be as high as 2000 W/cm^2 and such components cannot be cooled by any cooling technique other than immersion of such components in suitable dielectric liquid where heat is removed by boiling [6]. Even in forced convection, the heat dissipation capacity of air is much lower than that of water. The possible heat flux that can be met at various temperature differences by different cooling techniques is provided in **Figure 2**. Based on heat flux (exact or range) and maximum allowable temperature for the component or device, the selection of cooling technique can be made by using the chart given in **Figure 2**. Like, for a heat flux of 0.1 W/cm^2 and allowable temperature difference of 60°C, natural convection with radiation can be considered, whereas for a heat flux of 1 W/cm^2, forced convection will lead to more than 100°C temperature rise and so immersion with fluorocarbons can be considered [6]. In principle, for larger heat fluxes, either forced water cooling or immersion with boiling are the choices.

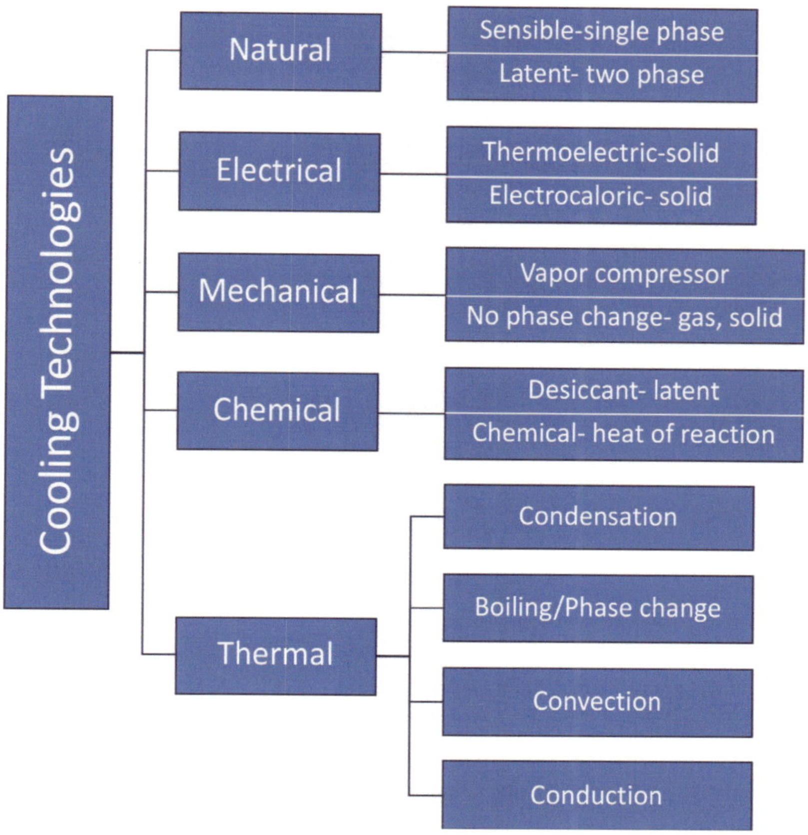

Figure 1. Schematic classification of cooling technologies based on energy input.

Recent review and analysis on the cooling technologies by the author revealed that microchannel-based forced convection and phase-change (like heat pipe) cooling are among the most promising techniques that are suitable and capable of removing very high heat loads [2]. However, because of very compact, lightweight, and superior convective cooling performance (for microscale hydraulic diameter of the channel), extensive research works have been performed on the application of microchannel-based cooling systems (e.g., heat sinks) in cooling high heat-generating devices [7–9]. Although heat pipes are already used commercially and have large market value, microchannel-based cooling technology is mostly in the development stage showing great prospect. Details about cooling technologies, their mechanisms, performance, and suitability for specific cooling need can be found in literatures [2, 5, 6] and will not be discussed here.

2.2. Emerging coolants

Although good progress has been made in the innovation and advancement of cooling techniques, very little advancement of coolants and coinage of new coolants have been made.

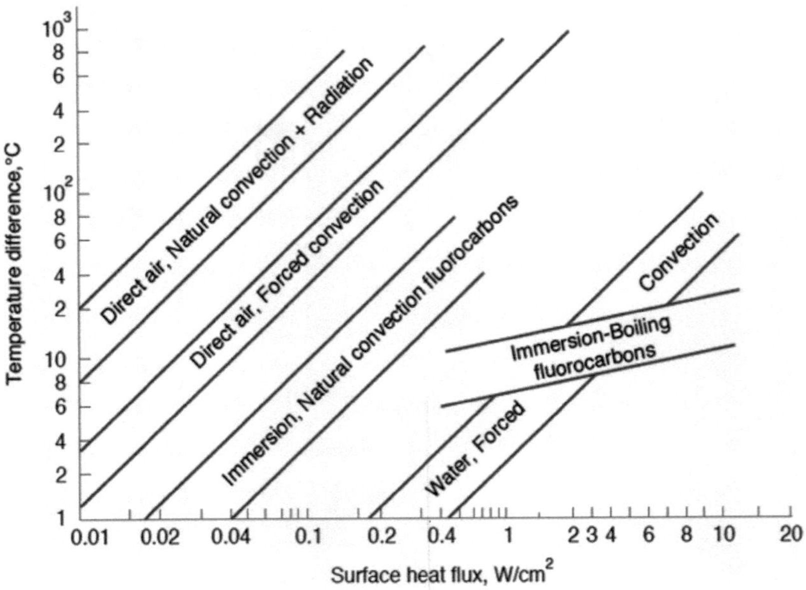

Figure 2. Cooling method for various heat flux and specified temperature differences [6].

This is mostly due to the limitation of improving cooling performance of conventional fluids without compromising other important factors. Nonetheless, there is always a need to have fluids with superior cooling capability compared to the conventional ones. Thus, researchers have long been trying to either develop new class of fluids or to improve the thermal properties of conventional heat transfer fluids.

Subsequently a new class of fluids termed—nanofluids, which are the suspensions of nanoparticles in conventional fluids, was coined in 1995 [10]. As a new type of engineered fluids (coolants), nanofluids exhibit mostly substantially higher heat transfer properties and features such as thermal conductivity, heat capacity, convective and boiling heat transfer compared to conventional fluids [11–15]. These affirm that nanofluids are better coolants compared to conventional fluids. In addition, number of research works (e.g., [3, 16–18]) have been made where nanofluids were directly employed in cooling systems of commercial electronic or computing devices to evaluate the cooling performance of these new fluids. The results of these studies were promising as nanofluids in those cooling systems resulted in better cooling performance as compared to conventional cooling fluids. However, despite great prospects and extensive research conducted on these new fluids, there are many controversial and inconclusive issues remained. Another new type of nanofluids named—ionanofluids, which are the suspensions of nanoparticles in ionic liquids, was recently devised by our group [19]. Primary findings showed that these new nanofluids also exhibit superior thermal properties and show potential compared to pure ionic liquids for cooling applications [19–22].

Nevertheless, these emerging liquid coolants still need to be carefully studied and assessed for their performance and sustainable applications in advanced cooling technologies.

3. Summarizing contributions

Except this short note, each chapter's contribution and findings have been briefly highlighted in this section.

In the first contribution, the authors Jin and coworkers presented a novel energy-saving and environmentally friendly air-conditioning method which is an indirect evaporative air cooling heat exchanger based on M-cycle. Besides experimental investigation on this heat exchange system, they developed a computational model to study the impacts of key parameters on its performance. The model showed a close agreement with the experimental findings. It was demonstrated that the inlet air conditions and the dimension of the air flow passages considerably influence the cooling effectiveness. They recommended that the air flow passages can be redesigned to create greater turbulence resulting in a larger heat transfer coefficient so that a lower thermal resistance can be achieved. Also, the geometry of this indirect evaporative heat exchanger with a smaller hydraulic diameter and a higher Nusselt number can positively impact its cooling performance.

Spray cooling is one of the effective cooling technologies for high heat flux removal requirement and this cooling technique from two aspects: the entire spray (spray level) and droplets (droplet level) is thoroughly reviewed by the authors Gao and Li. Here, cooling is achieved through different modes like convective heat transfer from the heated surface to the film flow, nucleate boiling, liquid conduction inside the film flow, and interfacial evaporation from the liquid film. For the spray level, they emphasized the cooling performance to spray property and some key properties involving spray characterization, nozzle positioning, phase change, and enhanced surface are also summarized. Discussion on the droplet level is focused on the impact of droplet flow on film flow, which is the key flow mechanism in spray cooling. Besides highlighting advantages and barriers of using spray cooling for engineering applications, recommendations for future work and on unresolved issues in spray cooling are also made.

Development and evaluation of a novel heat pump-based energy system particularly for cooling high-power LED lamps and also for waste heat recovery was demonstrated by Cen and coworkers. They proposed an integrated system for temperature control and heat recovery to operate in conjunction with an LED lamp. The integrated system, which consists of a heat pump in which the LED lamp itself operates as the heat pump evaporator, adopts an active method to simultaneously achieve waste heat recovery and LEDs cooling. A prototype of this energy system was developed, and experiments were conducted to determine the effect of several parameters, such as cooling water flow rate and LED power on the LED lead-frame temperature, compressor power consumption, and system performance. Findings of this study reveal that this novel cooling system can lead to significant energy savings.

Author Vinicius evaluated module architectures and units of photovoltaic (PV) cooling systems aiming to determine, select, and design a modular system which can be applied in a real-scale photovoltaic power plant (PVPP) to enhance the yields of electricity production. A detailed analysis of the regional climatic, geographic, and solar conditions, as well as construction, operational, and maintenance aspects of this modular architecture-based cooling system is also provided. There are three major types of photovoltaic cooling systems which are photovoltaic-thermal (PVT) liquid and air collectors, PV ventilated with heat recovery, and non-PVT systems. Due to higher rates of heat exchange between the cooling fluid and the PV module, it was concluded that the best design and arrangement of the cooling system are of the coil and multiple-channel types.

Boiling is one of the most leading cooling mechanisms, and a comprehensive experimental study on nucleate pool boiling performance of refrigerants R-134a and R-410A on a powder-coated heating surface is reported by Dewangan and coauthors. Effects of heat flux and coating parameters on boiling performance are determined, and results are very promising as the boiling heat transfer coefficient was found to enhance by 1.9 times that of the plain heating surface. Besides evaluating the performance, it also presents an empirical correlation for the prediction of boiling heat transfer coefficient of refrigerants.

Whether conventional or advanced, the ultimate goal of development of any cooling technique is to remove the heat from the devices, systems, or processes in real-world applications. Thus, last two contributions focused on specific and industrial applications of cooling technologies. Cryogen spray cooling (CSC), which is highly transient with short spurt duration (several tens of milliseconds), has successfully been implemented in laser dermatology such as the treatment of port wine stain. The contribution by Zhou and Chen presents a review of the progress of this CSC in laser surgery for the treatment of port wine stain, focusing on flashing spray mechanism, spray and thermal characteristics of droplets, dynamic heat transfer, and strategies of heat transfer enhancement. It is demonstrated that the cooling ability of CSC which can be increased through alternative cryogens, new nozzles, and hypobaric pressure method is essential to improve therapeutic outcome, especially for darkly pigmented human skin.

In the last contribution, water-based centrifugal compressor cooling systems for industrial applications is carefully studied and analyzed by Hanslik and Suess. They focused on natural refrigerants as R404A is no longer a viable option for commercial refrigeration applications. The analysis of the systems has shown that a self-contained unit with water-loop combined with a centrifugal compressor cooling system with water as refrigerant proves to make sense from an energetic point of view. The main factors that influence the efficiency are the evaporation and ambient temperatures and the process pressure of the compressor. A promising option is the combination of a carbon dioxide process with a chiller using water as refrigerant.

4. Conclusions

This chapter briefly discusses various aspects of cooling technologies and emerging coolants besides highlighting main research contribution and results from each contributed chapter of this book. It is demonstrated that selection of cooling technology needs to be based on various key factors and their combination like heat removal capacity, suitability, and sustainability. Besides new cooling technologies, innovation and development of noble coolants are also crucial to meet the ever-increasing cooling demand. This book covers a number of key topics related to advanced cooling approaches, their performance, and specific applications. Finally, the book is believed to be a useful reference source of information and guide to all class of people who are involved or interested in the areas of cooling technologies and their applications.

Acknowledgements

This work was supported by the Fundação para a Ciência e a Tecnologia (FCT), Portugal through grant—PTDC/NAN-MAT/29989/2017.

Author details

S M Sohel Murshed

Address all correspondence to: smurshed@tecnico.ulisboa.pt

Center for Innovation, Technology and Policy Research, Department of Mechanical Engineering, Instituto Superior Tecnico, University of Lisbon, Lisbon, Portugal

References

[1] Murshed SMS. Electronics Cooling. Rijeka: INTECH; 2015

[2] Murshed SMS, Nieto de Castro CA. A critical review of traditional and emerging techniques and fluids for electronics cooling. Renewable and Sustainable Energy Reviews. 2017;**78**:821-833

[3] Murshed SMS, Nieto de Castro CA. Nanoparticles-loaded fluids for cooling modern electronics. AIP Conf. Proc. 2018;**1980**(1):050015

[4] Harrabin R. Cooling technologies set to become red hot sector. BBC Technology Report. 27 May 2016 (https://www.bbc.com/news/technology-36398002)

[5] Brown JS, Domanski PA. Review of alternative cooling technologies. Applied Thermal Engineering. 2014;**64**:252-262

[6] Cengel YA. Heat Transfer: A Practical Approach. New York: McGraw–Hill; 2003

[7] Wei Y, Joshi YK. Stacked microchannel heat sinks for liquid cooling of microelectronic components. Journal of Electronic Packaging. 2004;**126**:60-66

[8] Hassan I, Phutthavong P, Abdelgawad M. Microchannel heat sinks: An overview of the state-of-the-art. Microscale Thermophysical Engineering. 2004;**8**:183-205

[9] Garimella SV, Singhal V, Liu D. On-chip thermal management with microchannel heat sinks and integrated micropumps. Proceedings of IEEE. 2006;**94**:1534-1548

[10] Choi SUS. Enhancing Thermal Conductivity of Fluids with Nanoparticles. Vol. 231. ASME FED; 1995. pp. 99-105

[11] Murshed SMS, Nieto de Castro CA. Nanofluids: Synthesis, Properties and Applications. New York: Nova Science Publishers; 2014

[12] Murshed SMS, Leong KC, Yang C. Thermophysical and electrokinetic properties of nanofluids—A critical review. Applied Thermal Engineering. 2008;**28**:2109-2125

[13] Murshed SMS, Nieto de Castro CA. Conduction and convection heat transfer characteristics of ethylene glycol based nanofluids—A review. Applied Energy. 2016;**184**:681-695

[14] Murshed SMS. Determination of effective specific heat of nanofluids. Journal of Experimental Nanoscience. 2011;**6**(5):539-546

[15] Murshed SMS, Nieto de Castro CA, Lourenço MJV, Lopes MLM, Santos FJV. A review of boiling and convective heat transfer with nanofluids. Renewable and Sustainable Energy Reviews. 2011;**15**:2342-2354

[16] Roberts NA, Walker DG. Convective performance of nanofluids in commercial electronics cooling systems. Applied Thermal Engineering. 2010;**30**(16):2499-2504

[17] Escher W, Brunschwiler T, Shalkevich N, Shalkevich A, Burgi T, Michel B, et al. On the cooling of electronics with nanofluids. Journal of Heat Transfer. 2011;**133**:051401

[18] Nazari M, Karami M, Ashouri M. Comparing the thermal performance of water, ethylene glycol, alumina and CNT nanofluids in CPU cooling: Experimental study. Experimental Thermal and Fluid Science. 2014;**57**:371-377

[19] Nieto de Castro CA, Lourenço MJV, Ribeiro APC, Langa E, Vieira SIC, Goodrich P, et al. Thermal properties of ionic liquids and ionanofluids of imidazolium and pyrrolidinium liquids. Journal of Chemical and Engineering Data. 2010;**55**(2):653-661

[20] Nieto de Castro CA, Murshed SMS, Lourenço MJV, Santos FJV, Lopes MLM, França JMP. Enhanced thermal conductivity and heat capacity of carbon nanotubes-ionanofluids. International Journal of Thermal Sciences. 2012;**62**:34-39

[21] França JMP, Vieira SIC, Lourenço MJV, Murshed SMS, Nieto de Castro CA. Thermal conductivity of $[C_4mim][(CF_3SO_2)_2N]$ and $[C_2mim][EtSO_4]$ and their ionanofluids with carbon nanotubes: Experiment and theory. Journal of Chemical & Engineering Data. 2013;**58**:467-476

[22] Minea AA, Murshed SMS. A review on development of ionic liquid based nanofluids and their heat transfer behavior. Renewable and Sustainable Energy Reviews. 2018;**91**:584-599

Chapter 2

Energy Efficient Indirect Evaporative Air Cooling

Xin Cui, Xiaohu Yang, Yanjun Sun,
Xiangzhao Meng and Liwen Jin

Additional information is available at the end of the chapter

http://dx.doi.org/10.5772/intechopen.79223

Abstract

An energy-saving and environmentally friendly air-conditioning method has been proposed. The key component is a novel indirect evaporative heat exchanger (IEHX) based on the M-cycle. In this design, the compact IEHX is able to produce sub-wet-bulb cooling and reduce the air temperature approaching dew-point temperature. This chapter aims to achieve a fundamental understanding of the novel IEHX. A numerical model has been developed and validated by comparing the simulated outlet air conditions against experimental data. The model showed a good agreement with the experimental findings. Employing the validated numerical model, we have theoretically investigated the heat and mass transfer behavior occurred in the IEHX. The detailed cooling process has been analyzed on the psychrometric chart. In addition, the effects of varying inlet conditions and airflow passage dimensions on the cooling efficiency have been studied. By analyzing the thermal performance of the IEHX, we have provided possible suggestions to improve the performance of the dew-point cooler and enable it to attain higher cooling effectiveness.

Keywords: air conditioning, indirect evaporative cooling, heat exchanger, numerical simulation, heat and mass transfer

1. Introduction

The evaporative cooling technique takes the advantage of water evaporation to achieve cooling effect. As a potential alternative to the conventional mechanical vapor compression system, it has drawn great attention for building cooling applications.

An indirect evaporative cooling system is able to produce the cool air without moist change. For a typical indirect evaporative heat exchanger (IEHX), the primary air (or product air) and the secondary air (or working air) flow in separate passages. The secondary air in wet channel acts as a heat sink by absorbing heat due to water evaporation.

The evaporative cooling system has the following advantages over the current mechanical vapor compression system [1–4]: (1) energy and cost savings; (2) reducing peak power requirements; (3) no CFCs; (4) reducing pollutant releases; and (5) easily incorporation with existing systems. On the other hand, the traditional IEHX shows the following limitations: (1) the air humidity increases in direct evaporative cooling systems resulting in uncomfortable indoor thermal environment for humans; (2) the indirect evaporative cooling system generally has a low cooling efficiency [5]; and (3) the theoretical ultimate cooled air temperature is the wet-bulb temperature.

To further enhance the cooling performance of conventional IEHX, research works have proposed a novel regenerative IEHX, which can cool the air below its wet-bulb temperature [1]. Since this design was proposed by Maisotsenko, the airflow arrangement is also named as M-cycle [2, 6]. Several studies have theoretically and experimentally investigated the dew-point evaporative cooling system [7–9]. This type of IEHX is able to branch part of the pre-cooled primary air into the wet channel [10, 11].

In order to investigate the cooling performance of a regenerative IEHX, analytical models have been developed based on modified effectiveness-NTU method [10, 12]. In addition, numerical simulations have been carried out to study the impact of key parameters on a counter-flow IEHX [13–15]. Anisimov and Pandelidis [16] presented a numerical study analyzing the indirect evaporative cooler with four different configurations. Zhao et al. [17] conducted a numerical study on a novel dew-point IEHX. According to the simulation results, the wet-bulb effectiveness of the IEHX was greatly influenced by the dimension of the airflow passages, the inlet air velocity, and working-to-intake-air ratio. The cooler achieved the highest wet-bulb effectiveness of 1.3 for a typical summer condition in the UK. Moshari and Heidarinejad [18] developed a numerical model and solved the governing equations by using finite difference method in MATLAB. The regenerative evaporative heat exchanger has been demonstrated to provide sub-wet-bulb cooling. Jradi and Riffat [19] conducted an experimental and numerical investigation on a dew-point evaporative cooling system. An optimized design of the indirect evaporative heat exchanger has been presented to obtain a dew-point effectiveness of 78%. Ham and Jeong [20] proposed a novel design on dew-point evaporative heat exchanger to address the issues of complex ventilation control and energy waste. Moshari et al. [21] investigated indirect evaporative cooling systems with one- and two-stage to analyze the optimum configuration by considering the cooling effectiveness, water consumption and thermal comfort. Chen et al. [22] developed an analytical model to study the heat and mass transfer process in an indirect evaporative cooler as a pre-cooling device in tropical areas. The model was established taking account of the condensation in the dry channels. Woods and Kozubal [23] conducted an experimental and numerical study on a desiccant-enhanced air conditioner by combining a dew-point evaporative heat exchanger and a desiccant dehumidifier. The second stage of this air conditioner was a counter-flow regenerative IEHX, which had a similar configuration compared with the IEHX designed by Zhao et al. [17] and Riangvilaikul and Kumar [24].

In this chapter, we first introduce the design of novel indirect evaporative heat exchangers, followed by the mathematical description, which was employed to study the air treatment process in the IEHX. An experimental study was then conducted to validate the computational model. Finally, the validated model was used to study the cooling performance of the novel dew-point IEHX and to investigate the impact of several influential parameters.

2. Description of novel indirect evaporative heat exchanger

In this section, we introduce two types of IEHX. The first type is a counter-flow regenerative IEHX. The other one is a novel dew-point IEHX based on the modification of the M-cycle arrangement.

The schematic of a one-unit channel pair of a typical counter-flow regenerative IEHX is shown in **Figure 1**. A number of this type of channel pairs is stacked to form the structure of the IEHX. The intake air flows in alternative primary channels (dry channels). Before the outlet of the dry channel, a part of the primary air is diverted into the secondary channel (wet channel). The primary air in the IEHX can be cooled along the flow passages without changing the humidity ratio.

Figure 2 illustrates the airflow in novel dew-point IEHX in terms of a one-unit channel pair. It comprises a product channel and adjacent working channels. By spraying water in working wet channels, inner surfaces are maintained in wet conditions. The working channel employs a closed-loop arrangement. The working air firstly enters the working dry channels, and it is then recirculated into the wet channels. The configuration is able to pre-cool the working air before entering the wet side. Therefore, at the recirculation point of the working channel, the air stream achieves a high cooling capacity as a result of a lower wet-bulb temperature compared with the inlet air. Theoretically, the outlet temperature of product air can be reduced toward its dew-point temperature.

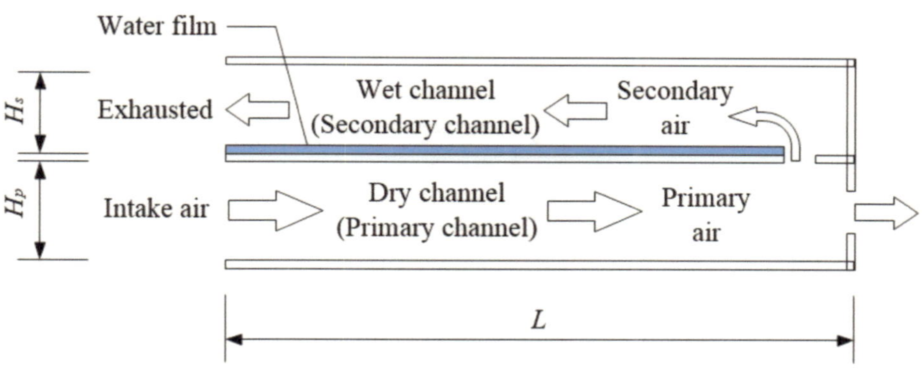

Figure 1. Schematic of a one-unit channel pair of a typical counter-flow regenerative indirect evaporative heat exchanger.

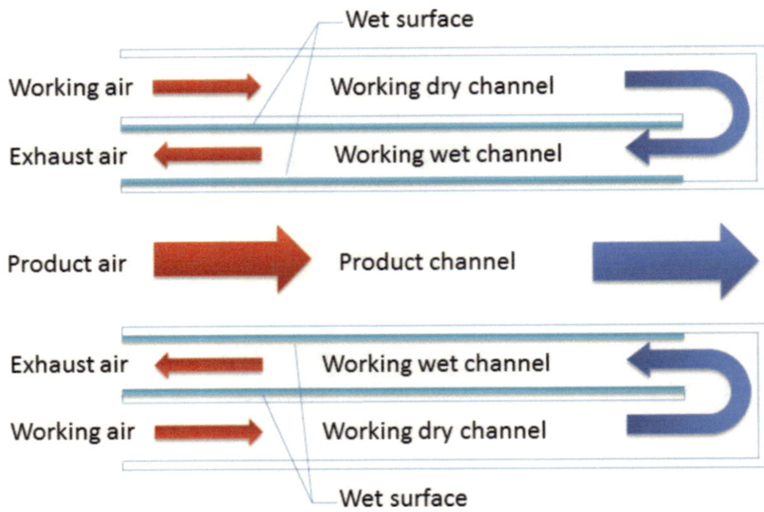

Figure 2. Schematic of the novel dew-point IEHX (one-unit channel pair).

3. Mathematical modeling

To develop a mathematical model predicting the air treatment performance in the IEHX, the following assumptions are considered: (1) the airflow is steady and incompressible; (2) the channel height is comparatively small; (3) the airflow is fully developed and laminar; (4) the surface of wet channel is covered with a thin water film; and (5) the outer surface is insulated.

The moist air flowing in both the working channel and the product channel is governed by the following equations [9].

The continuity equation:

$$\frac{\partial u_a}{\partial x} + \frac{\partial v_a}{\partial y} = 0 \tag{1}$$

Momentum conservation equation:

$$u_a \frac{\partial u_a}{\partial x} + v_a \frac{\partial u_a}{\partial y} = -\frac{1}{\rho_a}\frac{dp}{dx} + \nu_a \frac{\partial^2 u_a}{\partial y^2} \tag{2}$$

Energy conservation equation:

$$u_a \frac{\partial T_a}{\partial x} + v_a \frac{\partial T_a}{\partial y} = \alpha_a \frac{\partial^2 T_a}{\partial y^2} \tag{3}$$

Equation of species diffusion for water vapor is expressed as

$$u_a \frac{\partial c_a}{\partial x} + v_a \frac{\partial c_a}{\partial y} = D_a \frac{\partial^2 c_a}{\partial y^2} \tag{4}$$

The inlet boundary conditions of the air are indicated as

$$u_a = u_{a,in}, \quad v_a = 0, \quad T_a = T_{a,in}, \quad c_a = c_{a,in} \tag{5}$$

At the air-water interface, the moist air is assumed to be saturated with the water film temperature. The vapor pressure of the saturated moist air can be expressed as a function of temperature using the following equation [25]:

$$ln\, P_{sat} = \frac{C_1}{T_w} + C_2 + C_3 T_w + C_4 T_w^2 + C_5 T_w^3 + C_6 ln\, T_w \tag{6}$$

where C_1 = −5.800220 6 E3, C_2 = 1.391499 3, C_3 = −4.864023 9 E−2, C_4 = 4.176476 8 E−5, C_5 = −1.445209 3 E−8, C_6 = 6.545967 3, and T_w is the absolute temperature of water film.

The water vapor concentration is specified as

$$c_a = \frac{P_{sat}(T_w)}{RT_w} \tag{7}$$

where $P_{sat}(T_w)$ is the saturated vapor pressure at the absolute temperature of the water film. The evaporation rate of water is determined by the gradient diffusion as:

$$m_v = M_{H_2O} D_a \left(\frac{\partial c_a}{\partial y} \right)_w \tag{8}$$

As a result, the interfacial condition at the water film surface in the working wet channel is given as

$$u_a = 0, \quad v_a = 0 \tag{9}$$

$$-k_w \frac{dT_w}{dy} = -k_a \frac{dT_a}{dy} + M_{H_2O} h_{fg} D_a \frac{\partial c_a}{\partial y} \tag{10}$$

To evaluate the cooling efficiency of the indirect evaporative heat exchangers, following expressions (wet-bulb effectiveness and dew-point effectiveness) are defined.

$$\varepsilon_{wb} = \frac{T_{a,in} - T_{a,out}}{T_{a,in} - T_{wb,in}} \tag{11}$$

$$\varepsilon_{dew} = \frac{T_{a,in} - T_{a,out}}{T_{a,in} - T_{dew,in}} \tag{12}$$

4. Experimental study

A prototype of the plate-type counter-flow regenerative IEHX was fabricated as schematically illustrated in **Figure 3**. The one-unit channel pair comprises a product channel and two adjacent working channels. The intake air, which consists of both the working air and the product air, flows into the product channel (dry channel). At the end of the product channel, part of the product air is branched into the adjacent working wet channels by passing through the perforated plates. The dimensions for the counter-flow regenerative IEHX are listed in **Table 1**.

The schematic of the experimental setup for the IEHX is shown in **Figure 4**. A variable speed blower, which was used to control the intake airflow rate, was equipped at the inlet of the IEHX. The intake air temperature was adjusted by a heater before the blower. The measured parameters included the air temperature, humidity, and velocity.

Thermistors with accuracies of ±0.1°C were employed to measure the air dry-bulb temperature. The probes were inserted into the center of the airflow passages to measure the relative

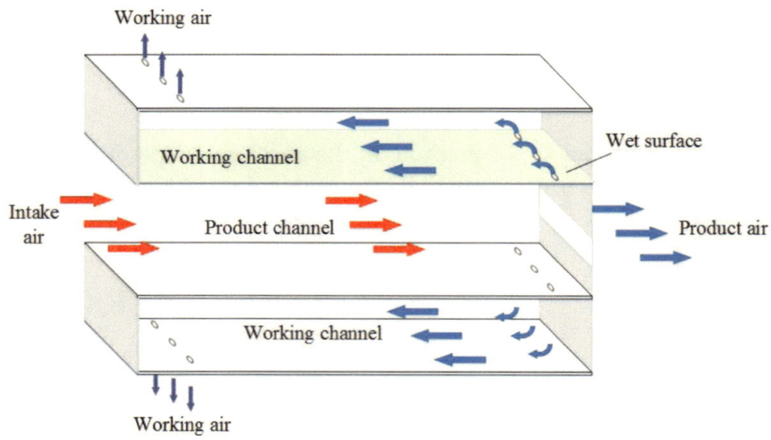

Figure 3. Schematic of the counter-flow regenerative IEHX.

Dimension	Symbol	Value	Units
Product channel gap	$H_{product}$	10	mm
Working channel gap	$H_{working}$	6	mm
Channel length	L	750	mm
Channel width	W	300	mm
Wall thickness	δ_p	0.3	mm
Wick thickness	δ_{wi}	0.2	mm

Table 1. Specifications for the counter-flow regenerative IEHX.

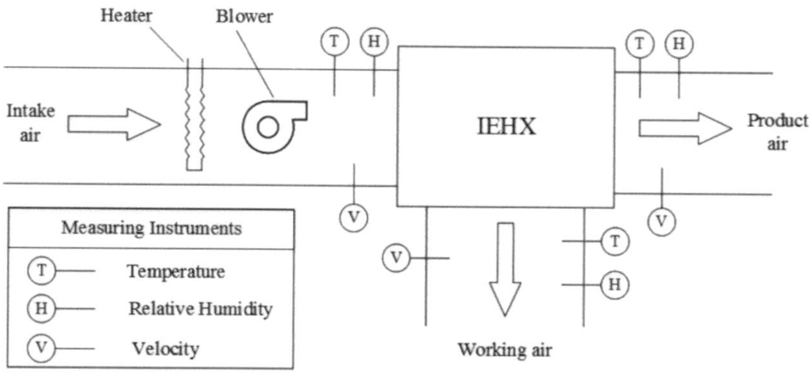

Figure 4. Schematic diagram of the experimental setup.

	Flow rate I	Flow rate II
Flow rate of intake air (L/s)	4.5	6
Velocity of product air (m/s)	1.5	2.0
Velocity of working air (m/s)	1.0	1.3
Inlet air humidity ratio (g/kg)	10	10
Inlet air temperature (°C)	22–29	22–29

Table 2. Operating condition of the counter-flow regenerative IEHX.

humidity and velocity of the air stream. The measured data were recorded using a data acquisition unit.

The experimental study was carried out to study the performance of the IEHX under varying inlet conditions. The inlet air velocity and temperature were adjusted to a stable condition by controlling the air blower and the heater. The operating conditions in this study were presented in **Table 2**.

5. Results and discussion

5.1. Model validation

The experimental data were first employed to validate the numerical model for the regenerative IEHX. The experimental condition was replicated in the simulation. **Figure 5** illustrates the comparison between the calculated outlet air temperature and the experimental results. The numerical model shows a good agreement within a maximum discrepancy about 5%.

The numerical model was further validated against experimental data presented by Woods and Kozubal [23, 26]. The experimental study was conducted for an M-cycle indirect evaporative heat exchanger. The supply air temperatures were measured under different operating

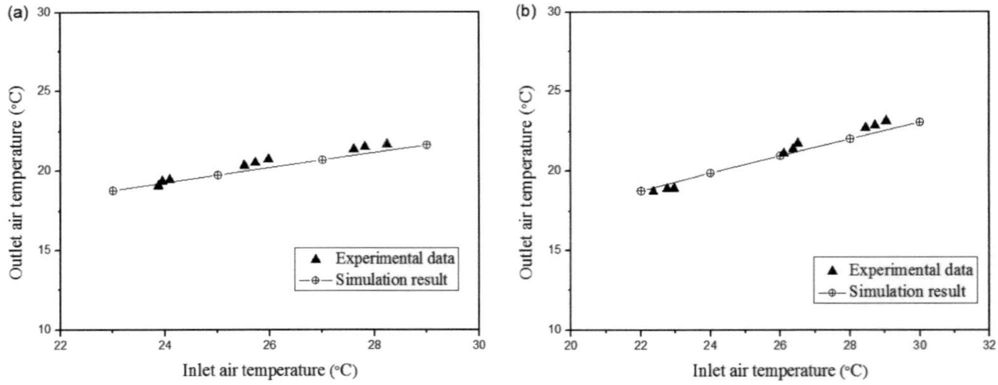

Figure 5. Validation 1: comparison between modeled results and experimental data on the counter-flow regenerative IEHX (a) flow rate of intake air is 4.5 L/s; (b) flow rate of intake air is 6 L/s.

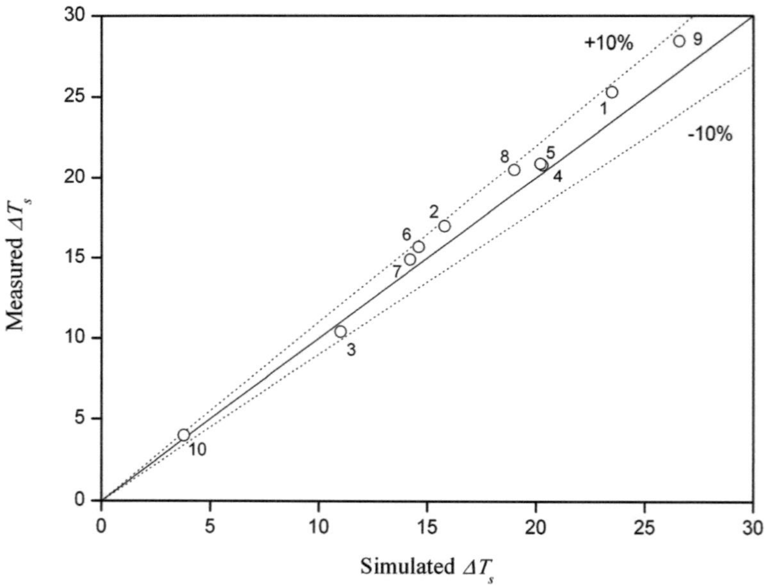

Figure 6. Validation 2: compare the supply air temperature change.

conditions. The validation was performed by comparing the simulated supply air temperature reduction with the experimental data. The discrepancy was within ±10% as shown in **Figure 6.**

By comparing the simulated results with experimental data, we can draw the conclusion that the validation has demonstrated the capability of the numerical model to theoretically investigate the cooling process for the IEHX.

5.2. Psychrometric analysis of the cooling process

By using the validated model, we investigated the air treatment process of the novel dew-point IEHX. Simulations were carried out to predict the states of air stream in the flow passages of the cooler [5]. The assumed inlet conditions were as follows: the dry-bulb temperature was 35°C and the humidity ratio was 10 g/kg for both working air and product air. The geometry parameter and the air velocity were maintained as the pre-set conditions as shown in **Table 3**.

The psychrometric analysis of the air conditions in the novel dew-point IEHX is shown in **Figure 7**. The inlet conditions for both the working air (point W1 in **Figure 7**) and the product

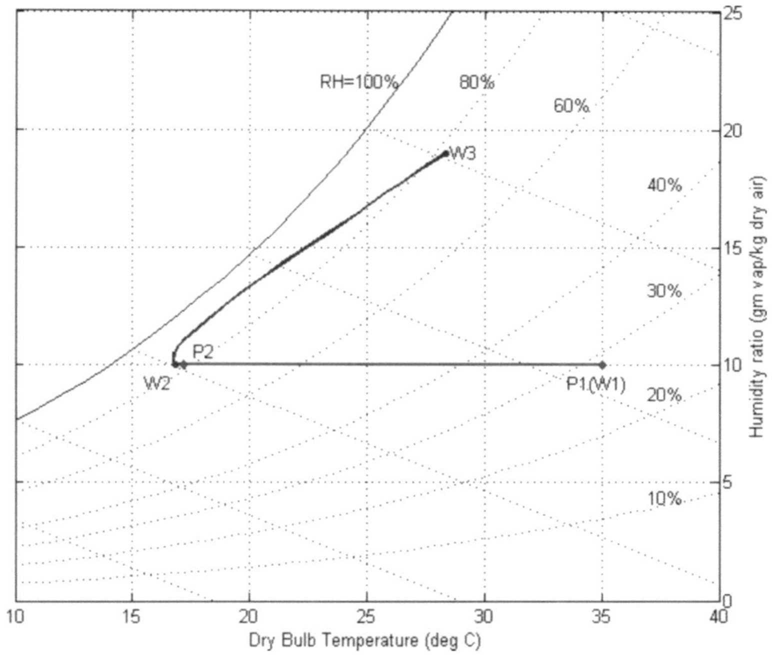

Figure 7. Illustration of air conditions on the psychrometric chart.

Parameters	Value	Units
L	1	m
$H_{product}$	6	mm
$H_{working}$	3	mm
$T_{air,in}$	35	°C
$\omega_{air,in}$	10	g/kg
$V_{product\ air}$	1	m/s
$V_{working\ air}$	1	m/s

Table 3. Pre-set conditions for the novel dew-point IEHX.

air (point P1 in **Figure 7**) are the same. In the product channel, the air dry-bulb temperature decreases with constant humidity ratio, resulting in a change of the condition from point P1 to point P2. In the working dry channel, sensible heat is transferred to the wet channel so that the working air temperature is reduced from state W1 to state W2. Since the air (at point W2 in **Figure 7**) is directed into the adjacent wet channel, it is humidified in the working wet channel where the heat is absorbed as a result of vaporizing water. The air in the working wet channel is finally exhausted at state W3.

Figure 8 illustrates the dry-bulb temperature profiles in airflow passages including one product channel and two working channels. In both the working dry channel and the product channel, the air temperature decreases in the direction of the airflow. In the working wet channel, the lowest air temperature is achieved. It can be inferred from **Figure 8** that the heat is transferred from dry channels to working wet channels. The working air absorbs heat in the working wet channel resulting in a temperature increase along the flow direction in the wet channel [11].

5.3. Influence of inlet air temperature and humidity

Figure 9 shows the wet-bulb effectiveness and dew-point effectiveness of the novel dew-point IEHX under different inlet air temperature and humidity conditions. In this study, the inlet air temperature varied in a range of values from 25 to 40°C and the humidity ratio was changed from 8 to 20 g/kg. The inlet air velocity and other geometry parameters were maintained as specified in **Table 3**.

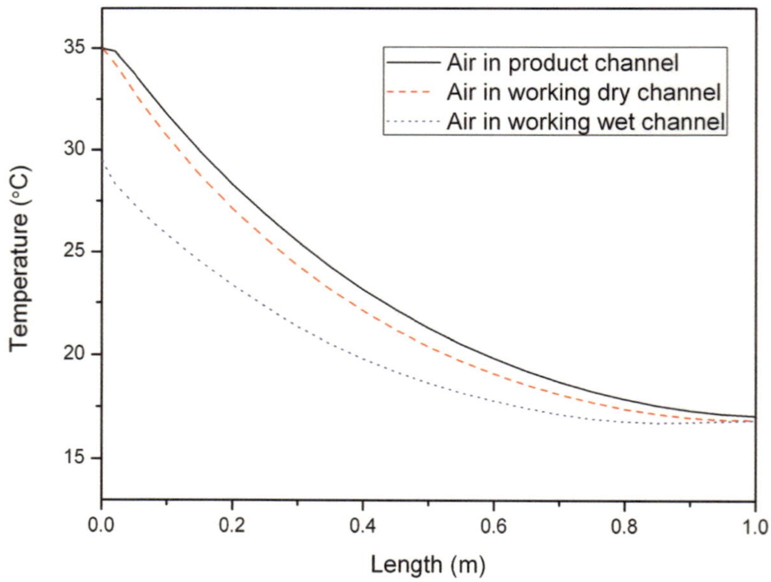

Figure 8. Temperature profiles of air in dry working channel, wet working channel, and product channel.

http://dx.doi.org/10.5772/intechopen.79223

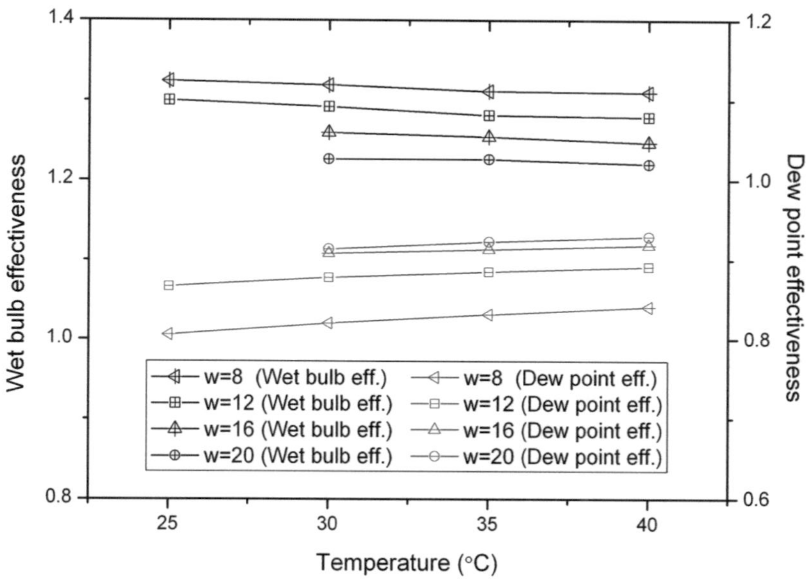

Figure 9. Cooling effectiveness for different inlet air conditions.

As shown in **Figure 9**, the dew-point effectiveness ranges 81–93%, and the wet-bulb effectiveness spans from 122 to 132%. Simulation results illustrate that the wet-bulb effectiveness of the IEHX is above 100%, which demonstrates the capability of the IEHX to produce air with a temperature lower than the wet-bulb temperature of inlet air.

When the inlet air humidity ratio is low, the vapor pressure gradient at the air-water interface is large resulting in a greater driving force for mass transfer. As a consequence, the working air is able to absorb more moisture during the process of vaporizing water, and the working air has a greater capacity to cool the air in adjacent dry channels. Therefore, a higher wet-bulb effectiveness can be reached by decreasing the inlet air humidity ratio.

Another finding from **Figure 9** is that the dew-point effectiveness may be decreased for a lower inlet air humidity ratio. It can be attributed to the following reason. As shown in the psychrometric chart, the dew-point temperature decreases significantly for a lower humidity ratio. Therefore, the temperature difference between the dew-point and the inlet dry bulb temperature shows a larger value at low humidity ratio [11].

5.4. Influence of the channel dimension

The cooling efficiency of the IEHX is also affected by the channel height and channel length. Simulations have been conducted by using two types of the channel dimensions. For the first type, the height of product channel ($H_{product}$) and the height of working channel ($H_{working}$) were assumed as 6 and 3 mm, respectively, while maintaining the channel length ranging from 300 to 1500 mm. In the second type, the height of product channel ($H_{product}$) and the height of working channel ($H_{working}$) were assumed as 10 and 5 mm, respectively, while maintaining the channel length spanning 300–2000 mm.

Figure 10 illustrates the correlation between the dimensionless channel length ($L/H_{working}$) and the wet-bulb effectiveness. It can be seen from the figure that the simulation results are consistent for the two sets of dimensions. A higher cooling efficiency can be achieved under a longer dimensionless channel length. In other words, increasing the channel length is able to improve the heat and mass transfer due to the increased contact area and time. Therefore, in order to achieve sub-wet-bulb cooling, the dimensionless channel length is suggested to be more than 200.

5.5. Thermal resistance analysis

Table 4 indicates the thermal resistances with examples values for the counter-flow regenerative IEHX. The channel dimension significantly influences the thermal resistances due to convective heat transfer (R_1 and R_4). In addition, the modified thermal resistance for convective heat transfer in wet channel (R_4) is impacted by the change of wet-bulb temperatures of the working air which also dominate the value of ξ ($\xi = \frac{\Delta h}{\Delta T_{wb}}$). The Nusselt number can be acquired from the available literature [27] for fully developed laminar flow. For instance, the Nusselt number of 8.234 is employed for the fully developed laminar flow with parallel walls.

The impact of channel height on the thermal resistances for convective heat transfer (R_1 and R_4) is shown in **Figure 11**. The thermal resistance for conductive heat transfer in water film and plate is associated with the material and its thickness. **Figure 12** presents the impact of thickness on the thermal resistances for conductive heat transfer (R_2 and R_3). The IEHX can be made from a variety of materials [28, 29]. As shown in **Figures 11** and **12**, the thermal resistance

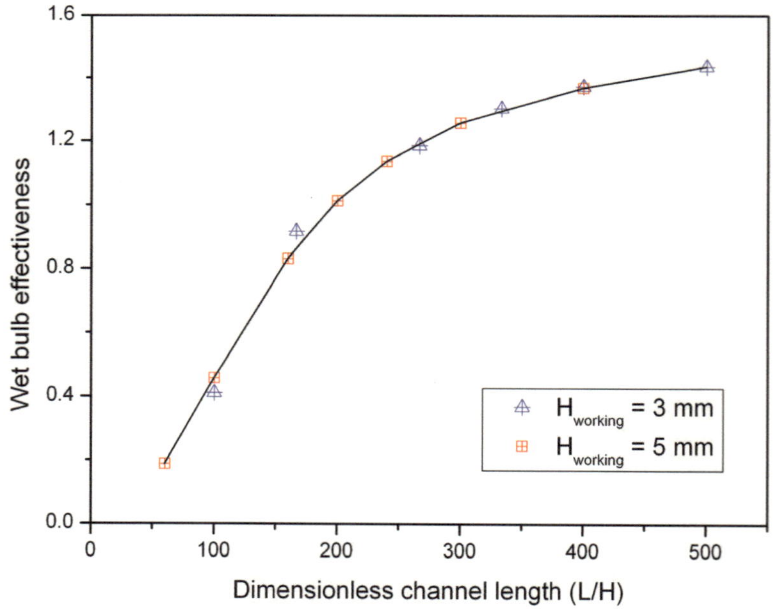

Figure 10. Impact of the dimensionless channel length on the cooling effectiveness.

Thermal resistance	Expression	Example values
R_1: thermal resistance for dry channel	$R_1 = \frac{1}{h_c} = \frac{D_1}{Nu \cdot k}$	$R_1 = 0.045821\ m^2\ K/W$, if $Nu = 8.235$ and $D_1 = 0.01\ m$
R_2: thermal resistance for plate	$R_2 = \frac{\delta_p}{k_p}$	$R_2 = 1.48\ E{-}6\ m^2K/W$, if δ_p =3 mm, and k_p =202 W/(m K)
R_3: thermal resistance for water film	$R_3 = \frac{\delta_w}{k_w}$	$R_3 = 4.92\ E{-}4\ m^2K/W$, if δ_w =3 mm
R_4: modified thermal resistance for wet channel	$R_4 = \frac{1}{\xi h_m} = \frac{c_{pa} D_2}{\xi Nuk}$	$R_4 = 0.013612\ m^2\ K/W$, if $Nu = 8.235$, $D_2 = 0.01$ m, and ξ =3.4 kJ/(kg·K)

Table 4. Thermal resistances in the IEHX.

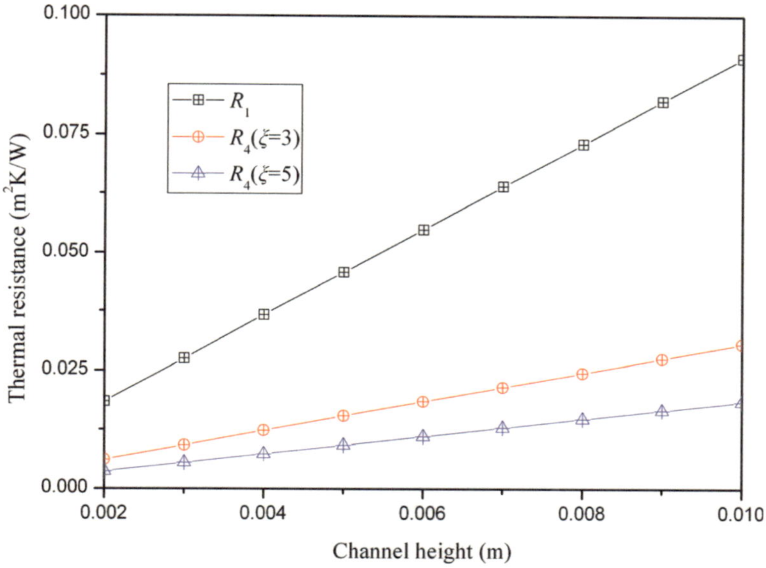

Figure 11. Thermal resistance for convective heat transfer under varying channel height.

R_2 is relatively small compared with the thermal resistance R_1 and R_4, especially when the plate has a small thickness. In general, the thermal resistances for convection heat transfer are much larger than the thermal resistances for conduction. Therefore, the thermal resistance R_2 and R_3 have marginal influence on the overall heat transfer of the whole system [7].

According to the analysis on thermal resistances, following suggestions are obtained in order to enhance the cooling performance of the IEHX. The airflow channels may be redesigned to generate larger flow turbulence resulting in a lower thermal resistance so that a larger heat transfer coefficient can be achieved [30]. The Reynolds number of the airflow in this IEHX is close to 1100. Therefore, the channel can be modified to create the flow turbulence by installing physical ribs in the airflow channels, which could potentially reduce the thermal resistance and increase the Reynolds number above 3000. In addition, the IEHX with a smaller hydraulic diameter will positively influence its cooling effectiveness due to a higher Nusselt number. To enhance the mass transfer process due to water evaporation in wet channels, some techniques may be employed involving increasing the water vapor concentration gradient between the water surface and the air stream.

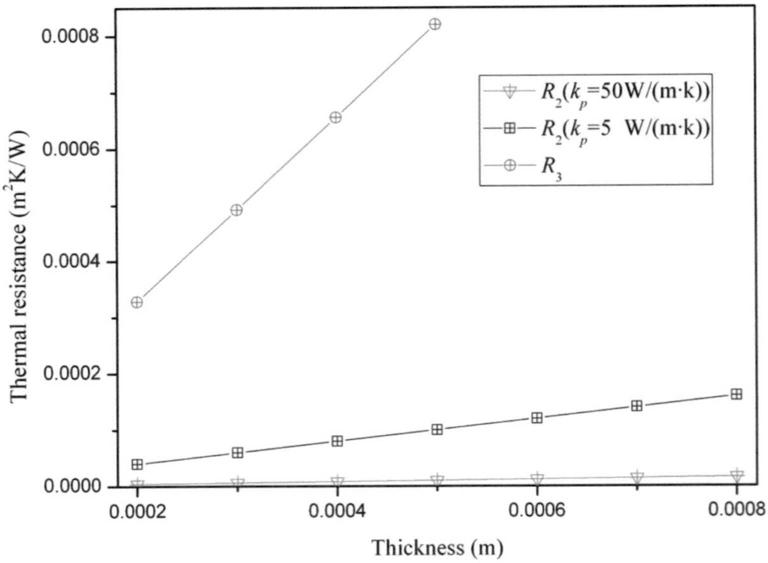

Figure 12. Thermal resistance for conduction in plate and water film.

6. Conclusions

The performance of novel indirect evaporative heat exchangers has been investigated. We developed a numerical model to study the influences of several parameters on the cooling performance. Wet-bulb and dew-point effectiveness were employed to evaluate the cooling efficiency of the IEHX. The cooling performance was impacted by the operating conditions and the structure of airflow passages. To promote cooling effectiveness, suggestions are provided to improve the thermal performance of the cooler. To achieve a lower thermal resistance, it is possible to modify the IEHX's channel and to create larger airflow turbulence. In addition, the performance of the IEHX can be positively improved by employing a geometry with a smaller hydraulic diameter.

Author details

Xin Cui, Xiaohu Yang, Yanjun Sun, Xiangzhao Meng and Liwen Jin*

*Address all correspondence to: lwjin@xjtu.edu.cn

Institute of Building Environment and Sustainable Technology, Shaanxi Engineering Research Center of Building Environment and Energy, School of Human Settlements and Civil Engineering, Xi'an Jiaotong University, Xi'an, Shaanxi, China

References

[1] Duan Z, Zhan C, Zhang X, Mustafa M, Zhao X, Alimohammadisagvand B, et al. Indirect evaporative cooling: Past, present and future potentials. Renewable and Sustainable Energy Reviews. 2012;**16**:6823-6850. DOI: 10.1016/j.rser.2012.07.007

[2] Chua KJ, Chou SK, Yang WM, Yan J. Achieving better energy-efficient air conditioning—A review of technologies and strategies. Applied Energy. 2013;**104**:87-104. DOI: 10.1016/j.apenergy.2012.10.037

[3] Jaber S, Ajib S. Evaporative cooling as an efficient system in Mediterranean region. Applied Thermal Engineering. 2011;**31**:2590-2596. DOI: 10.1016/j.applthermaleng.2011.04.026

[4] Caliskan H, Dincer I, Hepbasli A. A comparative study on energetic, exergetic and environmental performance assessments of novel M-cycle based air coolers for buildings. Energy Conversion and Management. 2012;**56**:69-79. DOI: 10.1016/j.enconman.2011.11.007

[5] Cui X, Chua KJ, Yang WM. Numerical simulation of a novel energy-efficient dew-point evaporative air cooler. Applied Energy. 2014;**136**:979-988. DOI: 10.1016/j.apenergy.2014.04.040

[6] Maisotsenko V, Gillan LE, Heaton TL, Gillan AD. Method and plate apparatus for dew point evaporative cooler. US Patent 6,508,402. 2003

[7] Cui X, Chua KJ, Islam MR, Yang WM. Fundamental formulation of a modified LMTD method to study indirect evaporative heat exchangers. Energy Conversion and Management. 2014;**88**:372-381. DOI: 10.1016/j.enconman.2014.08.056

[8] Cui X, Islam MR, Mohan B, Chua KJ. Developing a performance correlation for counter-flow regenerative indirect evaporative heat exchangers with experimental validation. Applied Thermal Engineering. 2016;**108**:774-784. DOI: 10.1016/j.applthermaleng.2016.07.189

[9] Cui X, Chua KJ, Islam MR, Ng KC. Performance evaluation of an indirect pre-cooling evaporative heat exchanger operating in hot and humid climate. Energy Conversion and Management. 2015;**102**:140-150. DOI: 10.1016/j.enconman.2015.02.025

[10] Hasan A. Going below the wet-bulb temperature by indirect evaporative cooling: Analysis using a modified ε-NTU method. Applied Energy. 2012;**89**:237-245. DOI: 10.1016/j.apenergy.2011.07.005

[11] Cui X, Chua KJ, Yang WM, Ng KC, Thu K, Nguyen VT. Studying the performance of an improved dew-point evaporative design for cooling application. Applied Thermal Engineering. 2014;**63**:624-633. DOI: 10.1016/j.applthermaleng.2013.11.070

[12] Liu Z, Allen W, Modera M. Simplified thermal modeling of indirect evaporative heat exchangers. HVAC&R Research. 2013;**19**:257-267. DOI: 10.1080/10789669.2013.763653

[13] Lee J, Choi B, D-YY L. Comparison of configurations for a compact regenerative evaporative cooler. International Journal of Heat and Mass Transfer. 2013;**65**:192-198. DOI: 10.1016/j.ijheatmasstransfer.2013.05.068

[14] Cui X, Islam MR, Mohan B, Chua KJ. Theoretical analysis of a liquid desiccant based indirect evaporative cooling system. Energy. 2016;**95**:303-312. DOI: 10.1016/j.energy.2015.12.032

[15] Cui X, Chua KJ, Yang WM. Use of indirect evaporative cooling as pre-cooling unit in humid tropical climate: An energy saving technique. Energy Procedia. 2014;**61**:176-179. DOI: 10.1016/j.egypro.2014.11.933

[16] Anisimov S, Pandelidis D. Theoretical study of the basic cycles for indirect evaporative air cooling. International Journal of Heat and Mass Transfer. 2015;**84**:974-989. DOI: 10.1016/j.ijheatmasstransfer.2015.01.087

[17] Zhao X, Li JM, Riffat SB. Numerical study of a novel counter-flow heat and mass exchanger for dew point evaporative cooling. Applied Thermal Engineering. 2008;**28**:1942-1951. DOI: 10.1016/j.applthermaleng.2007.12.006

[18] Moshari S, Heidarinejad G. Numerical study of regenerative evaporative coolers for sub-wet bulb cooling with cross- and counter-flow configuration. Applied Thermal Engineering. 2015;**89**:669-683. DOI: 10.1016/j.applthermaleng.2015.06.046

[19] Jradi M, Riffat S. Experimental and numerical investigation of a dew-point cooling system for thermal comfort in buildings. Applied Energy. 2014;**132**:524-535. DOI: 10.1016/j.apenergy.2014.07.040

[20] Ham S-W, Jeong J-W. DPHX (dew point evaporative heat exchanger): System design and performance analysis. Energy. 2016;**101**:132-145. DOI: 10.1016/j.energy.2016.02.019

[21] Moshari S, Heidarinejad G, Fathipour A. Numerical investigation of wet-bulb effectiveness and water consumption in one-and two-stage indirect evaporative coolers. Energy Conversion and Management. 2016;**108**:309-321. DOI: 10.1016/j.enconman.2015.11.022

[22] Chen Y, Yang H, Luo Y. Indirect evaporative cooler considering condensation from primary air: Model development and parameter analysis. Building and Environment. 2016;**95**:330-345. DOI: 10.1016/j.buildenv.2015.09.030

[23] Woods J, Kozubal E. A desiccant-enhanced evaporative air conditioner: Numerical model and experiments. Energy Conversion and Management. 2013;**65**:208-220. DOI: 10.1016/j.enconman.2012.08.007

[24] Riangvilaikul B, Kumar S. An experimental study of a novel dew point evaporative cooling system. Energy and Buildings. 2010;**42**:637-644. DOI: 10.1016/j.enbuild.2009.10.034

[25] ASHRAE. ASHRAE Handbook-Fundamentals. Atalanta, GA: American Society of Heating, Refrigerating and Air-Conditioning Engineers; 2009

[26] Kozubal E, Woods J, Judkoff R. Development and analysis of desiccant enhanced evaporative air conditioner prototype (No. NREL/TP-5500-54755). National Renewable Energy Laboratory (NREL); 2012

[27] Kays WM, Crawford ME, Weigand B. Convective Heat and Mass Transfer. 4th ed. New York: McGraw-Hill Higher Education; 2005

[28] Al-Sulaiman F. Evaluation of the performance of local fibers in evaporative cooling. Energy Conversion and Management. 2002;**43**:2267-2273. DOI: 10.1016/S0196-8904(01)00121-2

[29] Zhao X, Liu S, Riffat SB. Comparative study of heat and mass exchanging materials for indirect evaporative cooling systems. Building and Environment. 2008;**43**:1902-1911. DOI: 10.1016/j.buildenv.2007.11.009

[30] Alam T, Saini RP, Saini JS. Use of turbulators for heat transfer augmentation in an air duct—A review. Renewable Energy. 2014;**62**:689-715. DOI: 10.1016/j.renene.2013.08.024

Chapter 3

Spray Impingement Cooling: The State of the Art

Xuan Gao and Ri Li

Additional information is available at the end of the chapter

http://dx.doi.org/10.5772/intechopen.80256

Abstract

The cooling of a surface can be achieved by the impingement of spray, which is a free surface flow of droplets ejected from a spray nozzle. Spray cooling can provide uniform cooling and handle high heat fluxes in both single phase and two phases. In this chapter, spray cooling is reviewed from two aspects: the entire spray (spray level) and droplets (droplet level). The discussion on the spray level is focused on the spray cooling performance as a function of fluid properties, flow conditions, surface conditions, and nozzle positioning. The advantages and barriers of using spray cooling for engineering applications are summarized. The discussion on the droplet level is focused on the impact of droplet flow on film flow, which is the key flow mechanism in spray cooling. Droplet flow involves single droplet, droplet train (continuously droplets broke up from jet flow), and droplet burst (droplet groups affecting at a constant frequency), and local cooling enhancement due to droplet flow is discussed in details. Future work and unresolved issues in spray cooling are proposed.

Keywords: spray cooling, spray property, enhanced surface, droplet train, droplet burst, flowing film

1. Introduction

Thermal management becomes increasingly important and challenging as the increase of power/heat density is taking place in many engineering applications, products, and industrial sectors. One example is the electronics industry. Advances in semiconductor manufacturing technology create more compact integrated circuits for electric devices. The latest Fin Field Effect Transistor (FinFET) technology contributes to the reduction of fabrication node from 22 nm in the year of 2012 to the current 10 nm, and even to 5 nm in 2021. Using a 10 nm FinFET manufacturing process, Apple A11 chip could contain 4.3 billion transistors on a die of

~87 mm^2, which is 30% smaller than the last version A10. In addition, thermal design power of electric chips, the maximum amount of heat removal from the electric chips, shows an increasing trend. As heat power density continues to grow, heat removal, also referred to as thermal management, is important for maintaining the temperature to meet material and safety constraints. In turn, the development and maintenance of electric devices rely on how effectively the heat is dissipated from the devices. The choice of cooling technology is a complicated systems work in high-power electronic, not only for fitting in the heat removal requirement from low power density to high power density, but also for considering the cooling efficiency, power load, overall power consumption of the cooling subsystem, and the cost of cooling infrastructure. This chapter focuses on fundamental heat removal capacities of cooling technology.

Different cooling technologies vary in their heat removal capacities, which are summarized in **Figure 1**. For low heat flux removal requirement, air-cooling, which removes the heat from the hot surface by airflow, is widely applied. The cooling performance can be enhanced by expanding the surface area or increasing the flow of air over the surface. The first approach is known as air free convection, while the second approach is air-forced convection. In comparison with free convection, the fluid motion in forced convection is generated by external source, for enhancing the local convection. In computers, cooling fins are added to heat sink for expanding the surface area, while a fan is attached to the cooling fins to enhance air convection. Heat flux by forced air convection can reach ~35 W/cm^2 while only ~15 W/cm^2 by free air convection (see **Figure 1**). Due to the increase of power density, many micro-electronic and power electronic devices now are in the range of heat flux beyond the air cooling capacity. Effective liquid cooling solutions are needed for thermal management of the high-heat-flux devices.

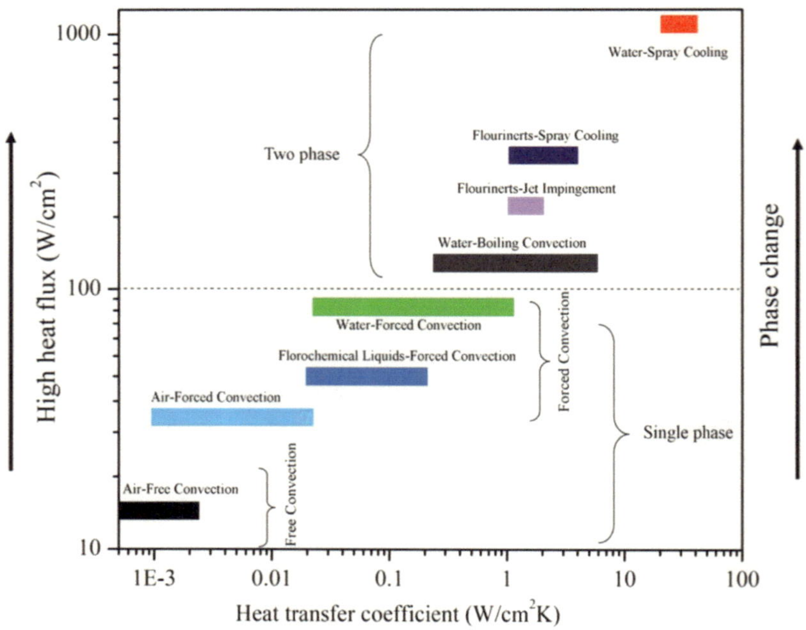

Figure 1. Heat removal capacity by applying different cooling technologies that is characterized by two parameters: Highest heat flux and heat transfer coefficient [1–4].

Spray cooling is one effective solution, which has the huge potential in handling the high heat fluxes in high-power electronics such as supercomputer, lasers, and radars. Spray cooling has several advantages over other cooling techniques. In comparison with air-cooling and jet impingement cooling, spray cooling owns a high heat flux removal capacity. Spray cooling can transfer heat in excess of 100 W/cm^2 using fluorinerts and more than 1000 W/cm^2 using water (see **Figure 1**). Due to high heat flux removal capacity, spray cooling allows precise temperature control with low fluid inventory [5]. Besides, spray cooling has uniform cooling temperature distribution over the entire spray-covered surface. This is because the entire spray-cooled area is receiving fresh liquid coolant droplets. For jet impingement cooling, the coolant flows radially outwards from the impingement spot. The radial flow has non-uniform temperature, and the largest subcooling and the optimal local cooling occur at the stagnation point. The non-uniform cooling results in non-uniform surface temperature in the cooling area, which could be significant for high heat fluxes.

However, there are still some barriers for applying spray cooling for engineering applications. Significant pumping power is needed to achieve large pressure drop through spray nozzle to produce fine spray, but the low cost is first priority in commercial application of cooling technologies. Another fact that the design and fabrication of spray nozzle do not follow the identical industry standard makes the unpredictable spray characterization. Hence, it is hard to get a universal correlation of spray characterization to cooling performance, which also limits the implementation of spray cooling. Additionally, in comparison with the jet nozzle, nozzle orifice through the spray coolant is even smaller, increasing the possibility of orifice clogging and the occurrence of the dry-out area on the heated surface [6]. In spite of these barriers, spray cooling is still a popular cooling technology and many successful applications were reported for supercomputer (CRAY X-1) [7], laser diode laser arrays [8], microwave source components [9] and NASA's reduced gravity aircraft [10].

In spray cooling, liquid coolant is emitted from a pressurized nozzle and breaks up into numerous droplets. The small droplets land on the cooled surface, where the flow of droplets becomes a thin liquid film radially flowing on the surface (see **Figure 2a**). The cooling is achieved through the convection heat transfer from the cooled surface to the film flow being impacted by continuous flow of droplets, nucleate boiling on the cooled surface, liquid conduction inside the film flow, and interfacial evaporation from the liquid film to the surrounding air. Spray cooling provides uniform cooling that can handle high heat fluxes in both single phase and two phases. The cooling performance as a function of spray characterization, flow conditions, surface conditions, and nozzle positioning was widely discussed in past decades. These studies focused on the relationship between the spray cooling performance and the entire spray flow. However, in these spray-level studies, the understanding of cooling mechanism of spray droplets is missing. At the droplet level, the impact conditions are classified into a few categories (see **Figure 2b**): (a) impact of single droplet on dry surface appearing in nucleate boiling, transition boiling and film boiling, (b) impact of single droplet on stationary film where the radial velocity of the film is close to zero, such as stagnation zone, (c) impact of single droplet on radially flowing film and (d) impact of droplet burst on flowing film (droplet groups that frequently impact the surface). Although spray impingement cannot be simply considered as the superposition of single droplets due to the interaction of the neighboring droplets [11], the study of local cooling

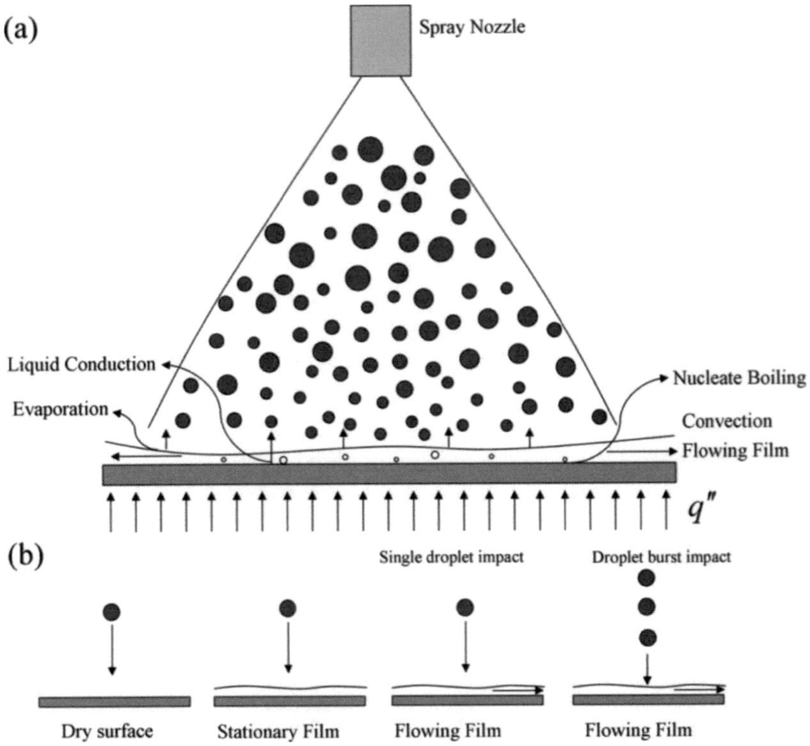

Figure 2. Spray cooling mechanism at the entire spray level (a) and droplet level (b).

performance at droplet level is still significant to the understanding of spray cooling mechanism, especially for the condition of the local film dominated by the droplet flow. Therefore, the research outcomes of spray cooling are reviewed from two aspects: the spray level and the droplet level.

2. Spray cooling at the spray level

Spray cooling can handle high heat flux in the constrictive space of electronic package when comparing to air-cooling, pool cooling, and jet cooling. This is because numerous fresh droplets generated by spray nozzle randomly affect the entire surface, and directly transfer the heat from surface to the coolant. The difference of fluid dynamics between spray impact and other cooling methods is a key factor affecting the mechanism of local heat transfer and resulting in different cooling performance. The first step of studying spray-cooling mechanism is to observe what happened on the heated surface. Numerous fundamental studies have been conducted theoretically and experimentally, which focus on the key parameters affecting impact dynamics and the relevant heat transfer mechanism. There are four aspects that have been demonstrated to significantly affect cooling performance, including spray characterization, nozzle positioning, phase change and enhanced surface [5, 6, 9].

2.1. Spray characterization

Since the earliest study on spray cooling by Toda [12, 13], many researchers put effort on spray characterization, the relevant cooling performance and the critical heat flux (CHF) in spray cooling. Spray characterization mainly involves droplet size, impact velocity, droplet flux, and volumetric flux. However, in experimental studies it is difficult to change only one parameter and isolate the remaining parameters. For example, on the cooled surface the increase of flow rate of coolant spray is accompanied with the increase of impact velocity and volumetric flux with a constant impact area. That is reason that the conclusions made on the dominant impact parameter are not consistent in previous studies of spray cooling.

Chen et al. [14] studied effects of three spray parameters of droplet size, droplet velocity and droplet flux on CHF. By adjusting spray nozzles, operating pressures, and spray distance between the nozzle exit and the heater surface, the effect of one spray parameter was studied while the others were kept constant. It was found that the mean droplet velocity is the most dominant parameter affecting CHF followed by the mean droplet flux, while the Sauter mean droplet diameter (d_{32}) has a negligible effect on CHF. In their later study [15], they further demonstrated the CHF varies with $N^{1/6}$ and $U^{1/4}$ (N and U are referred to as droplet flux and impact velocity). After determining local spray characteristics and local cooling in the water spray boiling curves, Mudawar and Valentine [16] found the dominant effect of the volumetric spray flux as compared to other spray parameters, and CHF was correlated to the volumetric spray flux and mean droplet diameter. After applying the dielectric coolant of PF-5052 with a lower boiling point of 50°C, Rybicki and Mudawar [4] concluded that the volumetric flux and droplet diameter are the most significant spray parameters influencing the spray cooling performance. Heat flux increases with increasing volumetric flux for a number of reasons. A larger fluid volumetric flux results in higher liquid velocity over the surface. The impact of the droplets onto the film can also interact with the liquid film, thinning the local thermal boundary layer.

In single-phase spray cooling, spray droplets land on a radially flowing film. Some researchers studied the property of the flowing film and its relation to spray cooling performance. Pautsch and Shedd [17] used a non-intrusive optical technique to measure the local film thickness generated by sprays. The film thickness was found to remain constant when the heat transfer mechanism was dominated by single-phase convection. Beyond the spray impact area, the dry-out phenomena appear even when the CHF is not reached. In the nucleate boiling regime, Horacek et al. [18, 19] measured the dry-out area, which was characterized by the three-phase contact line length, and measured using a Total Internal Reflectance technique. The wall heat flux was found to correlate very well with the contact line length. This contact line heat transfer mechanism was summarized by Kim [20] as one of main heat transfer mechanisms in the two-phase regime.

2.2. Nozzle positioning

Cooling performance can be influenced by changing the spray positioning. There are two significant positioning parameters in the study of spray cooling (see **Figure 3**): nozzle-surface distance H (the distance from the nozzle tip to the impact surface), and inclination angle θ (angle between spray axis and the surface normal). Geometrically, the spray impact on a

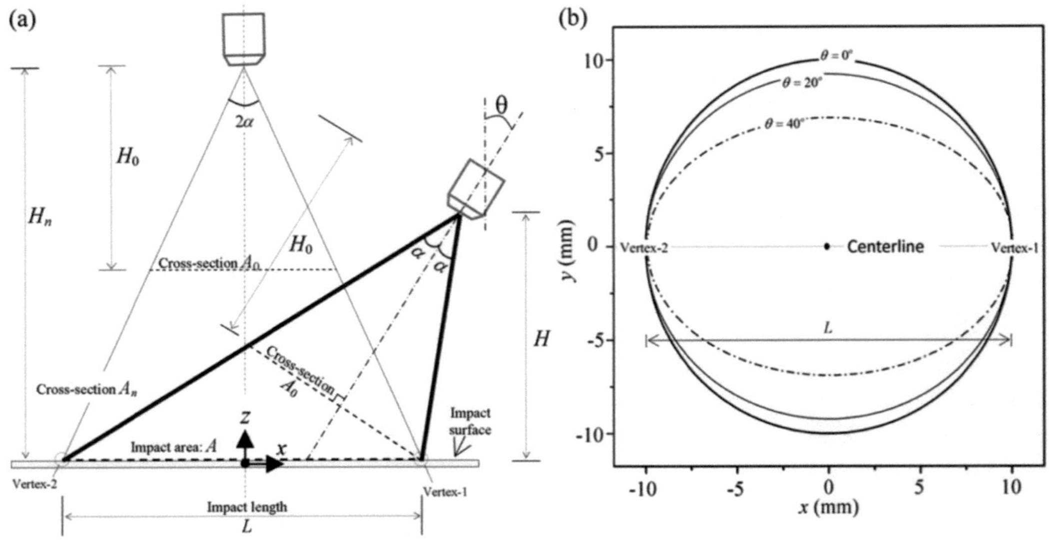

Figure 3. (a) The 2D geometry is on the central plane (z-x plane) of the cone perpendicular to the impacted surface (x-y plane). The positioning of the nozzle is determined by inclination angle θ and spray height H. H n is the required spray height by normal impact to cover a given impact length L. (b) Impact area with constant impact length L formed by the spray inclined at different angles θ [21].

surface is a spray cone insected by the impact surface. In normal spray impact changing nozzle-surface distance causes the change of spray covered area (see impact length L in **Figure 3a**), the local volumetric flux and film flow and relevant local cooling. Within the small surface area around 1 cm^2, Mudawar and Estes [22] found that the optimal nozzle-surface distance for the maximum CHF is achieved when the spray footprint is exactly enclosed within the cooling surface. Recently, in a larger surface area around 5.5 cm2 the optimal spray height for the largest area-averaged heat transfer coefficient was found to be smaller than that required for covering the entire heater area [21]. The reason is that the maximum local cooling appears at the edge of the impact, and the film flow outside the impact area still provides the effective cooling. However, this phenomenon is not significant in small heated area.

Some researchers focus on the effects of spray inclination on heat transfer performance. The impact area is circular for normal impact $\theta = 0^\circ$, and elliptical for inclined spray impact $0^\circ < \theta < (90^\circ - \alpha)$. The inclination angle is limited within $90^\circ - \alpha$. Otherwise, no spray lands on the cooled substrate. In experimental tests, the impact area, surface area and heater area are usually concentric, and the center is located at (x, y) = (0, 0) (see **Figure 3b**). Besides, impact length L maintains the same for studying the inclination influence on the cooling. Silk et al. [23] compared the cooling performance of three-inclination impact to normal impact on enhanced surface. It was found on both the flat surface and enhanced surface that heat flux increases with the increase of spray angle up to $\theta = 15°$. However, when $\theta > 15°$ the heat flux performance is little changed within the experimental uncertainty. Cooling enhancement by inclined spray is attributed to better liquid drainage through the elimination of the stagnation zone, which appears at the center in normal spray impact. Wang et al. [24] found that inclination of spray nozzle could enhance heat transfer if optimal orifice-surface distance is found. However, Visaria and Mudawar [25] indicated that inclination angle has minimal impact on the single-phase or two-phase regions of the boiling curve. Increasing inclination angle even decreases CHF and maximum CHF is always attained with the spray normal to the cooled surface. Rybicki and Mudawar [4] used upward-oriented and downward-oriented spray nozzles to assess their effects on cooling performance. The experimental results showed that the spray orientation has no measurable effects on the global cooling performance in both single-phase and two-phase regimes. Cheng et al. [26] found that the inclination angle would worsen heat transfer when the spray footprint is smaller than the heated surface. Therefore, the conclusions of these studies on spray inclination are contradictory.

There are three reasons addressed for contradictory conclusion of spray inclination. One is regarding the different nozzle positioning. As illustrated in **Figure 3**, two key parameters, spray distance and inclination angle, determine the nozzle positioning. However, at a certain inclination angle some studies [23] applied the constant spray distance, while others [4, 24, 25] adjusted the distance for the constant impact length. Another reason is related to the assumption of one dimensional steady-state conduction through the neck of cartridge heater for the surface heat flux calculation. Inclined spray impact causes considerable temperature difference on the cooled surface (see **Figure 4**). Hence, the radial conduction should be taken into account for inclined spray cooling. The last reason is from the surface temperature

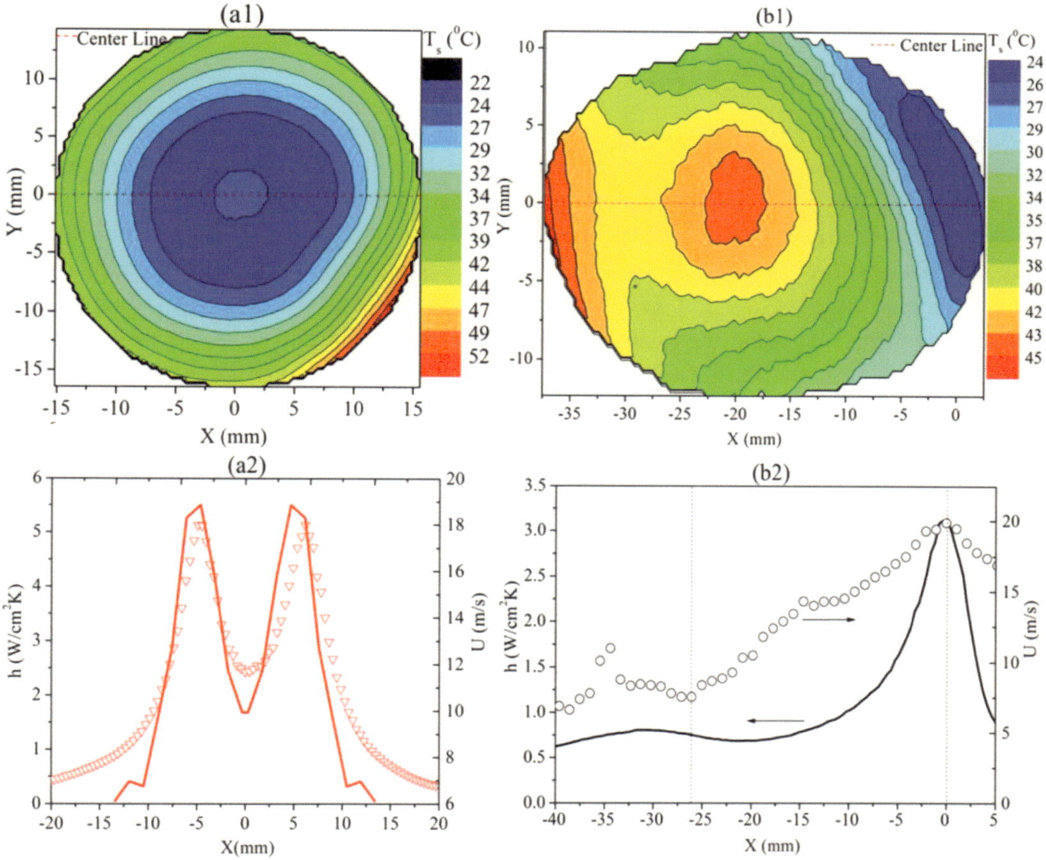

Figure 4. Local surface temperature distribution for normal impact (a1) and inclination affect with θ =30° (b1). Local droplet velocity and the relevant local heat transfer coefficient are plotted along the centerline in (a2) and (b2) [27].

measurement location. Different radial locations provide different temperature measurement due to significant temperature difference in inclined spray cooling.

To obtain surface temperature distribution in inclined spray, some researchers investigated local heat transfer by replacing cartridge heater with sputter-coated thin film heater, which enables infrared thermography for temperature measurement [21, 27, 28]. All of these studies found significant temperature difference on cooled surface for inclined spray cooling (one example in **Figure 4b1**). Gao and Li [27] compared the droplet impact velocity and heat transfer coefficient distribution along centerline for normal impact and inclined spray impact (see **Figure 4a2** and **b2**). The impact velocity was captured by a Stereo-Particle Imaging Velocimetry system. The trend line of heat transfer coefficient and droplet velocity shows clear correlation. For both cases, the locations of maximum droplet velocity coincide with the locations of the highest heat transfer coefficient. The further study by Gao and Li [21] indicated the global cooling shows slight diminishment for small inclination angle and enhancement for large inclination angles. On the central plane of the spray cone, the enhancement and diminishment of the local cooling performance are in general agreement with the increase and decrease of the spray flux. Thin film heater is not reliable for the surface temperature greater than boiling point, and experiments are tested in single-phase region. This is the limitation of thin film heater, and the robust heater for boiling test is needed for future study.

2.3. Phase change in spray cooling

Similar to pool boiling curve, the heat transfer curve of spray cooling can be separated to four regimes: single phase regime, nucleate boiling regime, transition boiling regime and film boiling regime [12, 13]. In the single phase regime, the heat flux linearly increases with increasing surface temperature difference between heater surface and coolant. Forced convection by radially moving film and evaporation on unsteady interface of thin film layer, play dominant roles in single-phase regime [29]. In the nucleate boiling regime, bubbles begin to repeatedly occur at nucleation sites on the heated surface, and the heat flux sharply increases as compared to single-phase cooling. Once the nucleation sites cover the heated surface completely, average heat flux will reach a peak value, which is defined as Critical Heat Flux (CHF).

Once reaching the CHF and coming to the transition boiling (decreasing region in the boiling curve), the efficiency of heat transfer on the heating surface significantly decreases. Liquid coolant absorbs heat from the surface and forms the vapor blanket, so the surrounding liquids are hard to get to the heater surface. That is the reason for the sharp decrease of heat flux in this regime. In the film, boiling regime an interesting phenomenon is an increasing trend of heat flux. Massive heat is generated from the heated surface and radiation heat transfer becomes a key heat transfer mechanism between the heated surface and the liquid, so the heat flux tends to increase from the Leidenfrost point. Considering the safety limit and fast implementation of electronic cooling, researchers' attention is paid to the theoretical correlation in single phase regime and nucleation boiling regime.

In the single phase regime, Rybicki and Mudawar [4] proposed the correlation for dielectric PF-5050 spray, which is

$$\mathrm{Nu} = 4.7\,\mathrm{Re}^{0.61}\,\mathrm{Pr}^{0.32} \tag{1}$$

Here Nu is the Nusselt number, Re is the Reynolds number, and Pr is the Prandtl number. Karwa et al. [30] developed a heat transfer correlation for full-cone water sprays, which is

$$\mathrm{Nu} = 20.344\,\mathrm{Re}^{0.659} \tag{2}$$

The correlation has an accuracy of ±7.3% for varied pressure drops. Heieh and Tien [29] studied R-134a spray cooling, and correlated the Nusselt number to the Weber number, size distribution and sensible heat effects in the single phase regime, which is

$$\mathrm{Nu} = 933\,\mathrm{We}^{0.36}\left(\mathrm{d}_{32}/d_0\right)^{0.25}\left(\Delta T/T_s\right)^{0.027} \tag{3}$$

In the nucleate boiling regime of spray cooling, the heat flux increases with the surface temperature faster than that in single-phase regime. Yang et al. [31] proposed two reasons. In nucleation, boiling bubble appears and grows on nucleation sites as the liquid coolant changes to the vapor. During the phase change, a larger amount of heat is removed from the heated surface, resulting in a temperature drop on nucleation sites. The other reason is attributed to the influence of secondary nucleation and evaporation on the heat flux enhancement [32]. When the numerous droplets impinge on heated surface, air is entrained into the liquid film, forming an air layer underneath the droplets. The air layer reaches the liquid-covered surface and finally breaks up into many tiny gas nuclei, which serve as secondary nucleation sites. Hence, the number of secondary nucleation sites is proportional to the droplet flux across the surface, which was proved in Yang's experiments [33]. Using water as coolant liquid, Mudawar and Valentine [16] proposed the CHF correlation with respect to the local volumetric flux Q'', and Sauter mean diameter (SMD) d32:

$$\frac{q''_m}{\rho_g Q'' h_{fg}} = f\left(\frac{\rho_f}{\rho_g}, \frac{\rho_f Q''^2 d_{32}}{\sigma}, \frac{\rho_f c_{p,f} \Delta T_{sub}}{\rho_g h_{fg}}\right) \tag{4}$$

In another study by Estes and Mudawar [34], a universal CHF correlation was constructed for spray cooling by using Fluorinerts FC-72 and FC-87 as well as the water.

$$\frac{q''_m}{\rho_g Q'' h_{fg}} = 2.3\left(\frac{\rho_f}{\rho_g}\right)^{0.3}\left(\frac{\rho_f Q''^2 d_{32}}{\sigma}\right)^{-0.35}\left(1 + 0.0019\frac{\rho_f c_{p,f} \Delta T_{sub}}{\rho_g h_{fg}}\right) \tag{5}$$

2.4. Enhanced surfaces

Enhancing spray cooling by changing the surface structure is one effective and low-cost approach, which benefits from optimal liquid management and enhancement of local cooling efficiency. According to the structure size, enhanced surface is classified into four categories: mini-structured surface, micro-structured surface, nano-structured surface, and hybrid-structured surface. Most of early studies of spray cooling have been conducted on flat surfaces. A few of them focus on the effects of surface roughness on cooling enhancement. Pais et al. [35] fabricated three rough surfaces using polishing grit with the size range of 0.3–22 μm and examined the roughness influence on heat removal capabilities. Tests showed that as the surface roughness decreases the CHF increases. CHF is up to 1200 W/cm² on the surface by polishing grit of 0.3 μm while only 1000 W/cm^2 on the surface by 22 μm grit. This is because the large surface roughness implies a thicker film thickness, leading to the later bubble breakup and departure, the impeding of vapor escape, the increased resistance to heat flux through evaporation on film surface, and the dampening of droplet impingement.

Mini-textured surfaces feature structure size above 1 mm, and the structure types of cubic pin fins, pyramids, and straight fins and so on (see **Figure 5a**). Silk et al. [23] observed that addition of finned structure to cooled surface decreases the convective thermal resistance, and increases the convection heat transfer relative to the flat surface, since the total wetted surface area is larger on the enhanced surface. Although the cubic pin fins and straight fins have the same wetted surface area, cooling performance of straight fins surface exceeds that of the cubic pin fins surface. This is attributed to liquid management on the heated surface and cooling efficiency on the wetted surface area. Xie et al. [39] indicated that the fin arrangement is a dominant factor in enhancing heat transfer rather than the wetted surface area. The improper fin arrangement causes the thick and slow moving liquid film and thus worsens the local cooling performance. This point of view needs further validation by measuring the change of local surface temperature.

Micro/nano or hybrid structured surfaces have been attracted huge attention to spray cooling as micro fabrication technology advances new micro-/nano-engineered surface in the last decade (see **Figure 5 b, c, d**). The experimental studies [36, 39–41] applied micro-textured surfaces with surface feature size from 25 to 480 μm, which is close to liquid film thickness but larger than average droplet size. Micro-textured surfaces showed slight effect on heat transfer enhancement in the flooded region, but greatly enhancing cooling performance in the thin film and partial dry-out regions as compared to the flat surface. The study by Zhang et al. [37] showed that nanostructured surface has better cooling performance since the contact angle is

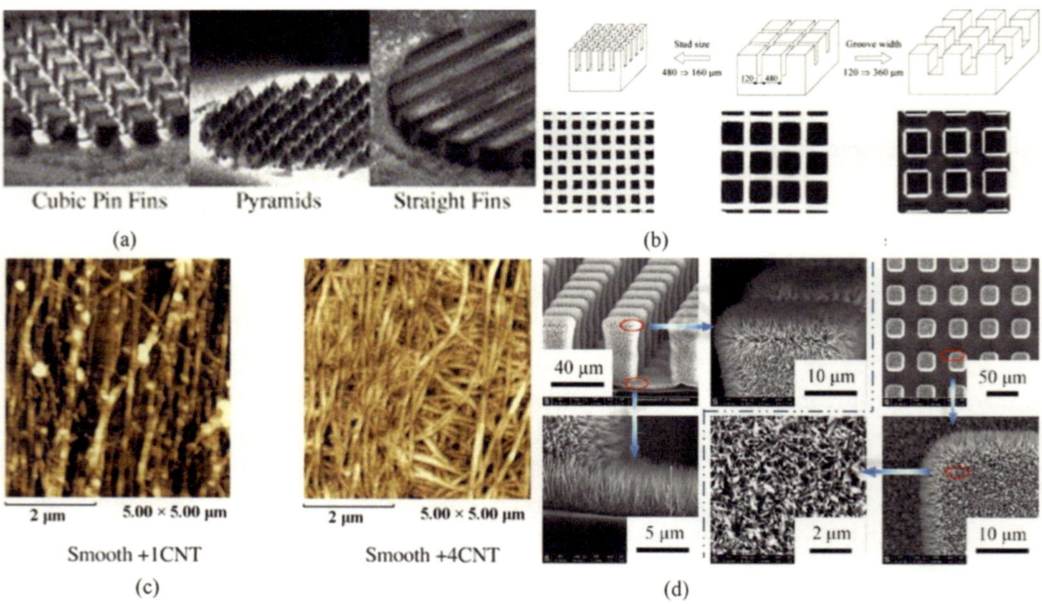

Figure 5. (a) Millimetric structured surface [23], (b) micro-structured surface [36], (c) Nano-structured surface [37], (d) hybrid micro/nano structured surface [38].

smallest on the nanostructured surface as compared to micro-structured surfaces and flat surfaces. Recently, Chen et al. [38] developed a hybrid micro/nano structured surface by growing the ZnO nanowire arrays on the top of etched micro-structured silicon wafer. Test results showed that cooling performance of hybrid surface is better than the micro-structured surface in boiling regime because of its great wetting capacity and reduction in dry-out surface area. If comparing performance of nanostructured surface [37] and hybrid surface [38], there is no significant difference in heat flux enhancement relative to the smooth surface.

3. Spray cooling at the droplet level

The impact dynamics during spray cooling is complicated as it involves many liquid phenomena, such as spreading, receding, splashing, droplet collision, generation of stationary film and radially flowing film, and liquid flooding. All of these impact phenomena result from the interaction of droplet flow and film flow on the impact surface. Droplet flow includes three types: single droplet, droplet train (continuous droplets formed from jet breakup), and droplet burst (portion of droplet train selected at a certain frequency). Similarly, film flow conditions involve dry surface (no film), stationary film, radially flowing film, or their combination on the cooling surface (see **Figure 6**).

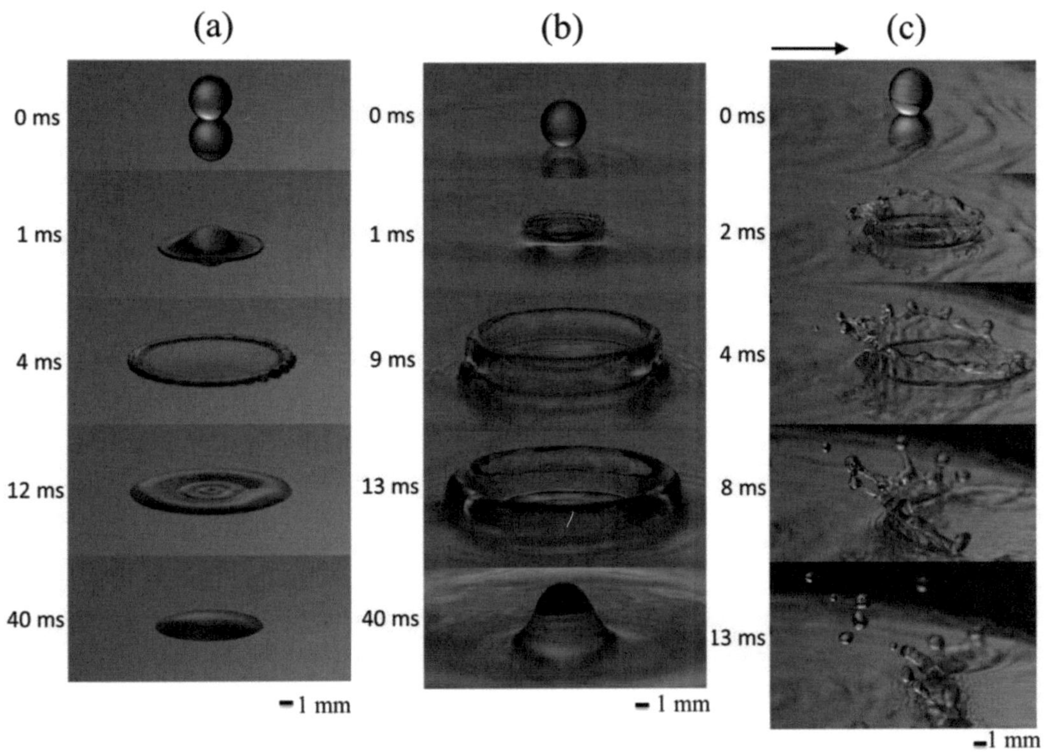

Figure 6. Single water droplets with same velocity and diameter (U 0= 1.85 m/s, D 0=3.2 mm) impact three different surface conditions: (a) dry surface, (b) stationary water film, (c) flowing water film [42].

The droplet and film flow conditions are two flow parameters directly determining the heat transfer mechanism of spray cooling. Coolant droplets bring significant temperature difference between the expanding droplet flow and flowing film, which contributes to the reduction of thermal resistance inside the film layer and enhancement of heat transfer from the heated surface to the flowing flow. Fluid dynamics on the impact surface is responsible for the local convection heat transfer. The fast flowing film transfers more heat to downstream. Thin film thickness reduces the thermal boundary layer and encourages evaporation from the liquid interface. Therefore, the fluid dynamics study of droplet affecting film enables us to get insight into thermal results of droplet impact on the film-cooled hot surface, and further understand spray cooling performance. The relevant literature is reviewed based on the droplet flow condition: single droplet impact, droplet train impact, and droplet burst impact.

3.1. Single droplet impact cooling

3.1.1. Impact on dry surfaces

The dry surface usually appears in two-phase spray cooling, which is shown by the change of contact line length. The researchers reported that the critical heat flux in spray cooling is achieved at the greatest contact line length. On dry surface, droplet impact dynamics on droplet-covered surface area is essential to local cooling performance. The process of a liquid droplet impact was divided by Rioboo et al. [43] into five successive phases: kinematic, spreading, relaxation, wetting, and equilibrium. Most research work has been focused on spreading and relaxation. In the spreading phase, contact line expands radially until reaching a maximum spreading, which is determined by droplet initial diameter, impact velocity, surface tension, viscosity, and wettability of the solid surface (Li et al. [44]). The maximum spread diameter is of critical importance in spreading phase. Clanet et al. [45] found that on a super-hydrophobic surface the maximal spread is significantly dependent on the viscosity of liquid droplets and scales as a function of Weber number~$We^{1/4}$. van Dam and Clerc [46] found a significant difference of maximum spread between substrates with small and large contact angles, showing the significant influence of wettability in the later stage of impact. A lower air pressure was found to suppress the droplet spreading, leading to a smaller maximum spread [47].

Some analytical models were proposed to predict impact process, most of which were based on the energy conservation of the impact droplet. Chandra and Avedisian [48] developed an empirical correlation of viscous dissipation, including estimated spreading time, simplified dissipation function, and estimated volume of viscous dissipation. Gao and Li [49] proposed a theoretical model based on the actual dynamic shape of the droplet that could successfully predict the maximum spreading diameter and receding diameter during the recoiling process. Some of the researchers put efforts on the investigation of splash using varied dry surfaces. Surface roughness and textures were demonstrated to influence the splash limit [50, 51]. Droplet impact on a moving surface was found to show different splash and non-splash phenomena as compared to stationary surfaces [52]. Previous studies on splash threshold under different surface conditions are summarized in **Table 1**.

Surface conditions	Threshold parameter K	Critical value K_c	References
Dry surface	$(WeRe^{1/2})^{1/2}$	57.7	Mundo et al. [50]
	$We^{0.5}Re^{-0.391}$	0.8458	Vander Wal et al. [53]
Moving dry surface	$WeRe^{1/2}(1-2.5\frac{U_s}{U_0}Re^{-1/2})^2$	5700	Bird et al. [52]
Stationary liquid film	$(WeRe^{1/2})^{0.8}$	2100	Cossali et al. [54]
	$We^{0.5}Re^{0.17}$	63	Vander Wal et al. [53]
Flowing liquid film	$WeRe^{1/2}(1+\bar{h}_f\bar{U}_f^2)(1+\bar{h}_f\bar{U}_f)^{1/2}$	3378	Gao and Li [42]

Table 1. Summary of splash thresholds under different surface conditions [42].

On heated dry surface, Bernardin et al. [55] mapped the boiling curve of droplet impact cooling as the same as the spray cooling. In the regime of single-phase liquid cooling, Pasandideh-Fard et al. [56] observed that increasing impact velocity would enhance heat flux around the impact area. This is because the raising droplet velocity promotes droplet spreading, thus increasing the wetted area on the heated substrate. However, increasing droplet impact velocity slightly enhances heat flux at the impact point. Batzdorf et al. [57] proposed a theoretical mode to predict the heat transfer rate during the droplet impact. The theoretical prediction is more accurate when the liquid Prandtal number $Pr>5$, since the droplet evaporation is not taken into account in the model. The predicted heat transfer rate shows a quick increase to the peak value and then the slow decreasing.

On superheated surface with temperature over 200°C, Tran et al. [58] found three significant phenomena after droplet impact: contact boiling (droplet contacts with the surface), film boiling (vapor layer formed underneath the droplet), and spray film boiling (vapor layer and tiny droplets ejected upward) (see **Figure 7a**). Their experiments showed that the maximum spreading of a droplet impact follows a universal scaling with the Weber number ($\sim We^{2/5}$), which is steeper than that on nonheated surface ($\sim We^{1/4}$) [45]. The steeper curve on heated surface results from a driving mechanism, which is caused by the evaporating vapor radially expanding and pushing liquid outward. Staat et al. [59] indicated that the Leidenfrost transition temperature shows little dependence on the Weber number of affecting droplet, but the transition to splashing shows a strong dependence on the surface temperature. Adera et al. [60] reported the formation of non-wetting droplets on a super-hydrophilic micro-structured surface by slightly heating the surface above the saturation temperature of the droplet fluid, which is contributed by the increased thermal conductivity and decreased vapor permeability of the structured region. In experimental study of Jung et al. [61], the transient temperature distribution during droplet spread was detected using infrared thermography. In contact boiling, the droplet coolant contacts the surface and the maximum heat flux is quick to reach at early impact stage ~2 ms at impact point. In film boiling, non-wetting surface appears at the early impact, and the maximum heat flux is even lower than that in contact boiling due to the existence of vapor layer underneath the droplet. On heated surface, the study of simultaneous impact of multiple droplets is few, which needs further discussion of droplet collision influence on contact line and local evaporation. This benefits the understanding of two-phase spray cooling and optimization of cooling efficiency.

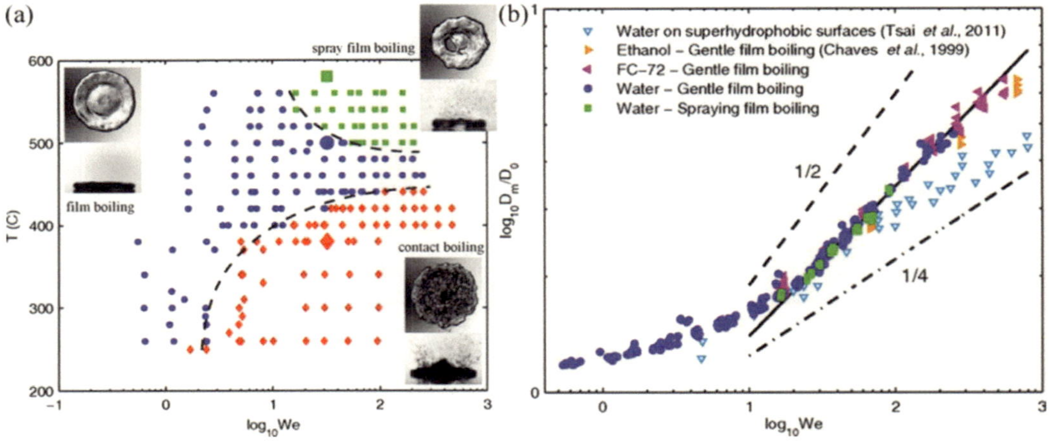

Figure 7. (a) Phase diagram of water droplet impact on a superheated surface [58], (b) plot of the maximum spreading diameter versus weber number [58].

3.1.2. Impact on stationary films

Stationary film occurs in the center of normal spray impact, or locates where the spray nozzle axis insects with the impact surface in inclined spray (see **Figure 2**). On a stationary film, most researchers focused on spread process and splash formation mechanism after impact. Yarin and Weiss [62] developed a quasi-one-dimensional model, which predicts the existence of a kinematic discontinuity in the velocity and film thickness distribution. The discontinuity corresponds to the emergence of an uprising liquid sheet. Roisman and Tropea [63] generalized Yarin's theory for the case of arbitrary velocity vectors in the liquid films both inside and outside the crown. Yarin and Weiss [62] experimentally found the crown radius from the impact center could be expressed as a function of the non-dimensional spreading time. Two empirical parameters existing in their model was given by the later study of Cossali et al. [64]. Droplet impact on a stationary film may or may not result in the splash. Finding the threshold condition for splash impact has been the focus of a few experimental studies. Cossali et al. [64] tested drops of various mixtures of water and glycerol affecting a thin liquid film and proposed an empirical parameter for predicting the occurrence of splash impact. For thick films, Cossali et al. [54] and Rioboo et al. [65] found a critical value of the threshold parameter, i.e. K_c=2100, above which splash impact occurs (see **Table 1**). To the authors' acknowledge few study has been conducted on heat transfer of single droplet impact on heated stationary film. For very thick stationary film, it likes a pool and the relevant heat transfer mechanism can be found in the study of pool boiling.

3.1.3. Impact on flowing films

The interaction between droplet flow and film flow is fundamental fluid dynamics in single-phase spray cooling or nucleate boiling (see **Figure 2b**). Impact dynamics was addressed in some researches. Alghoul et al. [66] presented an experimental investigation of a liquid droplet affecting onto horizontal moving liquid films. An asymmetrical crown shape was observed due to the effect of the moving film. Che et al. [67] demonstrated the on inclined falling flow asymmetrical crown shape is also formed after droplet impact. Gao and Li [42] further analyzed the early evolvement of droplet impact based on experiments and theoretical model (see **Figure 6c**). Once droplet lands on the film, the droplet flow quickly spreads and pushes the liquid outwards, causing the uprising liquid sheets. However, crown sheet is asymmetric owing to the collision mechanism on crown base. At the early stage of droplet impact, the direction of spreading flow is opposite to that of film flow at the upstream of impact point, while their direction is the same at the downstream. Uprising crown sheet may splash, which is dependent of the instability of the sheet rim. The stretching rate of crown sheet is a key factor influencing the rim instability. Analysis was conducted to derive equation of stretching rate, finding that the highest stretching rate appears at the location which droplet spreading flow is right opposite to the film flow, and the location is also the most probable location of splash. The value of splash threshold was provided to estimate whether splash occurs or not. The secondary droplets from splash fly away from the cooled surface, which do not contribute to the cooling performance. In other words, suppression of splash occurrence should benefit cooling enhancement.

The late study of Gao and Li [68, 69] further observed the whole development of droplet impact on flowing film, and demonstrated its relation to the local cooling. The impact process is observed by high-speed video, showing two states: spreading state, replacing state. In spreading state, the droplet flow spreads and gradually slows down until reaching the maximum spread. After that, the droplet flow is pushed towards the downstream and eventually replaced by the film flow. The measured temperature also shows two stages: response stage when the temperature quickly decreases, and recovery stage in which the temperature recovers to the steady state. An enhancement factor was proposed to indicate convection enhancement relative to the steady-state cooling. The peak enhancement is used to consider enhancement influence of impact velocity, droplet size and film flow rate, which is proportional to the square root of the ratio of the droplet flow rate to the film flow rate~$(U_0 D_0/Q')^{1/2}$. However, the conclusion made cannot be directly applied to spray cooling. One reason is that the film flow was generated by external source rather than by droplet flow itself. Another reason is that averaged cooling performance around the impact area was not involved.

3.2. Droplet train impact cooling

One possible phenomenon in spray cooling is that fresh droplets continuously impact the surface at a certain frequency. The droplet flow is defined as the droplet train flow. The fluid dynamics behind this is the interaction of continuous droplet train flow with the flowing film formed on the heated surface. To investigate heat transfer of spray cooling from this aspect, a few studies have been conducted on the heat transfer of continuous droplet train impinging on hot surfaces. Qiu et al. [70] demonstrated surface temperature influence on the impact dynamics. Prior to the steady state, the droplet film spreads on the heated surface, and the surface temperature enhances the spreading rate of the flowing film when the surface temperature is over the boiling point. With the increase of the surface temperature the steady-state film-wetted area decreases, and eventually maintains constant after the temperature is greater than 190°C. Besides, the temperature also affects the splashing angle (see **Figure 8**). A stable splashing angle marked by red line is established at higher surface temperature greater than 192°C. The later study of Qiu et al. [71] showed that the inclination of the droplet train decreases the splashing angle and increases the averaged secondary droplet size.

Soriano et al. [72] presented an experimental observation of multiple droplet train impingement. Impact spacing between multiple droplet streams would affect spreading and splashing in impact regimes, and the optimal cooling performance was achieved when the film velocity was not disturbed by adjacent droplet streams. Zhang et al. [73, 74] further demonstrated that both impact spacing and impingement pattern significantly affect local and global cooling performance on the hot surface. In comparison with the circular jet impingement cooling, the droplet train impingement achieves a better cooling performance for various impingement patterns. The same conclusion was made when comparing the cooling performance of droplet train and jet impingement on flowing film that cools the hot surface [75]. Through piezoelectric nozzles more groups of jet flow were generated and broke up to droplet train for cooling the hot surface [76], and the maximum heat flux reaches~ 170 W/cm^2 with the nozzle diameter of 25 μm. However, unclear impact dynamics and its relation to local cooling need the further study.

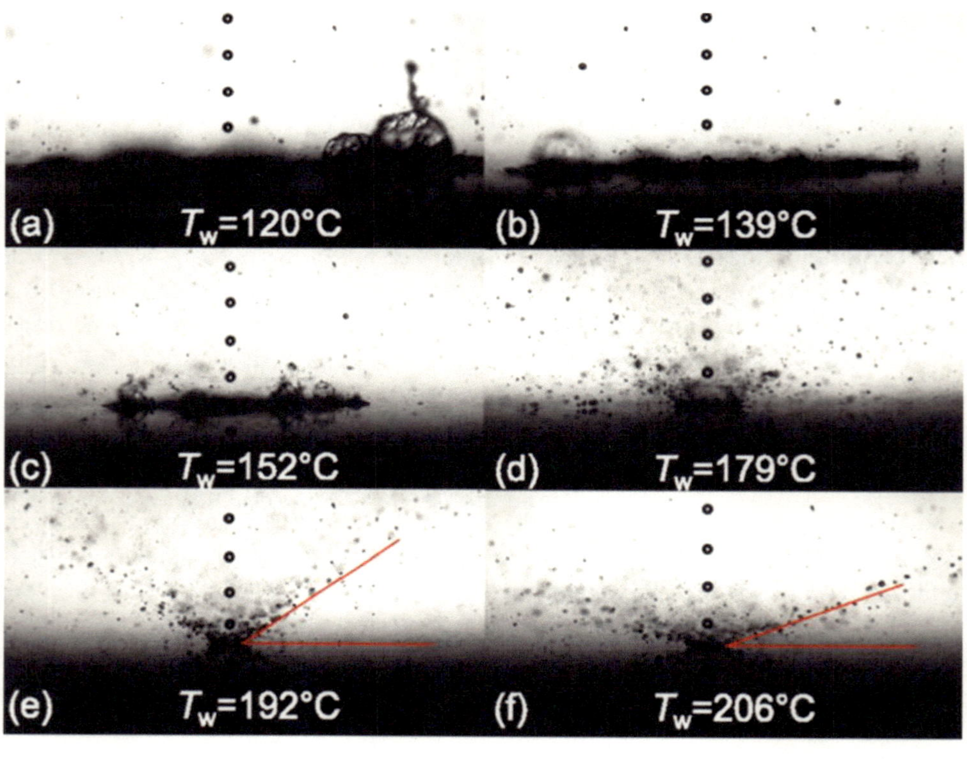

Figure 8. The impact dynamics of droplet train at different surface temperature and the droplet velocity is 15.2 m/s [70].

3.3. Droplet burst impact cooling

Our recent studies try to understand spray cooling from droplet burst aspect [75, 77]. Different from droplet train cooling, it assumed that in spray cooling droplet groups impact the surface at a constant frequency rather than droplet train. Each droplet group is defined as a droplet burst, and each burst contains a constant number of droplets, which is called burst size. The frequency at which droplet bursts are generated is called the burst frequency. The generation mechanism of droplet burst was first proposed by Gao and Li [75, 77] and implemented in tests. A droplet generator combined with controlled interrupter is applied for droplet burst generation. A droplet train is ejected from droplet generator with droplet frequency f_0, and a circular sector (aluminum plate) with a central angle θ serves as the interrupter. The droplet burst flow is generated by periodically interrupting a droplet train flow into a flow of droplet groups (see **Figure 9**). For each rotating of interrupter, there is only one droplet burst through the interrupter, so that the burst frequency is equal to the rotating frequency of interrupter f_b. The burst size n is determined by the burst frequency and central angle of interrupter $n = f_0 (1 - \theta/2\pi)/f_b$. For example in **Figure 9** (e), n = 6 when f_0= 1000 Hz, f_b= 13.5 Hz, θ = 330°.

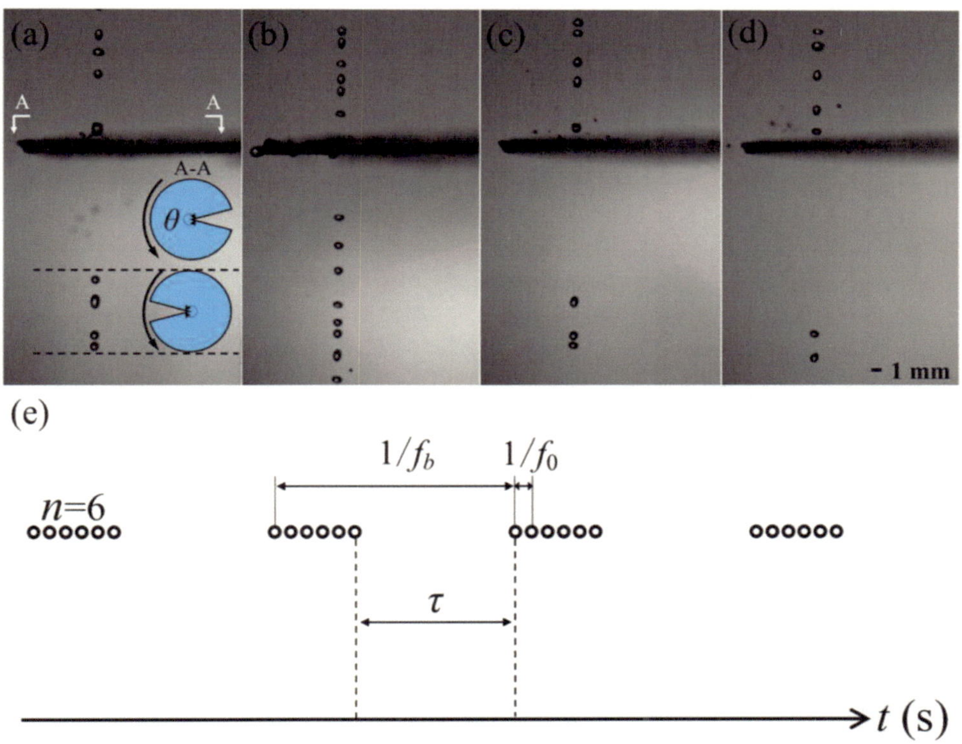

Figure 9. Droplet burst flows are generated by interrupting a droplet train flow (f 0 =1000 Hz) using an interrupter with an angle θ = 330° and varied frequencies f b: (a) 18.3 Hz, (b) 13.5 Hz, (c) 25.0 Hz, (d) 30.1 Hz, and (e) schematic of a droplet burst flow with n =6 [75, 77].

For the impact of one droplet burst (see **Figure 10**), at t = 0 s the local film flow has completely recovered from the impact of the previous droplet burst, but the temperature around the impact area is still lower than the film cooling temperature, showing residual effect from the previous droplet burst. As shown by the changes of T_s and h in **Figure 10** (b) and (c), the cooling enhancement is growing in extent and expanding in area at the early stage of droplet burst impact.

For the impact of one droplet burst flow, the temperature at impact point is measured. Temperature measurement shows that the burst flow causes the temperature to quickly decrease, and then the temperature fluctuates with the constant fluctuation frequency and amplitude in full-developed stage. The fluctuation frequency is equal to the burst impact frequency. The temperature at the impact point remains lower than the film cooling temperature without droplet burst impact. Heat transfer coefficient shows three development stages of the convection: affecting, restoring, and restored. During the restored stage, local cooling has returned to the film cooling. The restored stage may not exist if the time interval between bursts τ, is short (see **Figure 9**). This is because prior to reaching the restored stage the next droplet burst is coming, and local convection heat transfer goes to the next affecting stage.

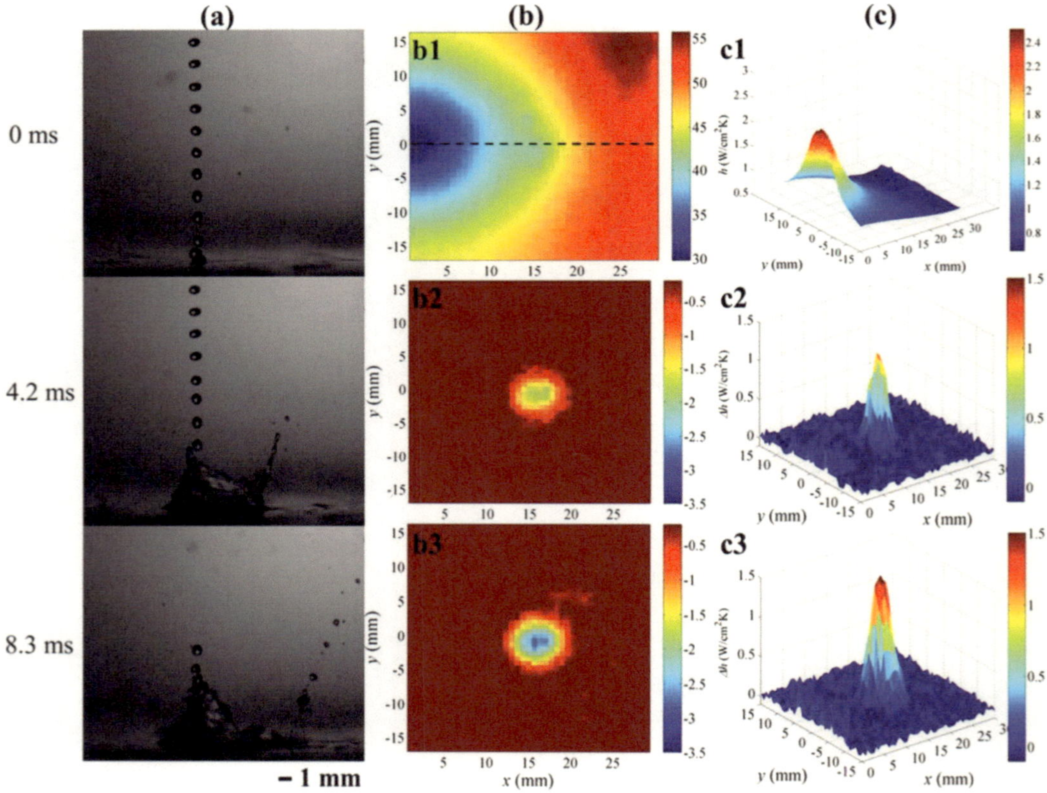

Figure 10. (a) Impact dynamics of a drop burst flow affecting the film flow; (b1) surface temperature distribution at t = 0 s; (b2) & (b3) temperature change; (c1) heat transfer coefficient at t = 0 s; (c2) & (c3) change of heat transfer coefficient [75, 77].

The comparison of burst flows shows that the trough value of the fluctuating temperature, T_{min}, decreases with increasing n. The temperature fluctuation amplitude, $T_{max} - T_{min}$, is dependent of burst size n and the time interval τ. The mean temperature is related to the number flow rate of the burst flow, indicating the number of droplets landing on the surface per second. The mean temperature decreases with increasing the number flow rate. Increasing the droplet impact velocity leads to the formation of stronger rising liquid sheets during the impact process. Significant increase of impact velocity reduces cooling enhancement due to the local loss of coolant caused by the rising liquid sheets and splashing. These conclusions made based on droplet burst cooling are good for the understanding of droplet characterization influence on cooling performance in spray cooling. The volumetric flux and droplet velocity are coupled in spray impingement. The larger volumetric flux is accompanied by the higher impact velocity. Determining which one is the dominant parameter is not reasonable in spray cooling performance. Observation of droplet burst cooling shows that the larger volumetric flux and proper impact velocity bring the better cooling enhancement.

4. Conclusion

Spray cooling is one effective cooling technology for handling high-power density and high heat flux removal requirement. In spray cooling, liquid coolant is emitted from a pressurized nozzle and breaks up into numerous secondary droplets affecting heated surface that is covered by radially flowing film. The cooling is achieved through the convection heat transfer from the heated surface to the film flow, nucleate boiling, liquid conduction inside the film flow, and interfacial evaporation from the liquid film. Based on research outcomes reported in the literature, spray cooling technology is reviewed from two aspects: the spray level and the droplet level. In the spray level, these studies emphasize the cooling performance to spray property. Some key properties are summarized in this chapter, involving spray characterization, nozzle positioning, phase change, and enhanced surface. In the droplet level, the studies focus on local heat transfer associated with droplet impact conditions, which are classified into a few categories: impact of single droplet on dry surface, stationary film, flowing film, impact of droplet train, and impact of droplet burst. Although spray impact cannot be simply considered as the superposition of single droplets, the studies in droplet level provide experimental and theoretical basis to explain what happened on heated surface and the relevant local heat transfer mechanism in spray cooling.

Author details

Xuan Gao and Ri Li*

*Address all correspondence to: sunny.li@ubc.ca

School of Engineering, The University of British Columbia, Kelowna, BC, Canada

References

[1] Glassman BS. Spray cooling for land, sea, air and space based applications, a fluid management system for multiple nozzle spray cooling and a guide to high heat flux heater design [Thesis]. University of Central Florida; 2005

[2] Sienski K, Eden R, Schaefer D. 3-D electronic interconnect packaging. In: Proceedings of IEEE Aerospace Applications Conference; 10 February 1998; Aspen, CO, USA. DOI: 10.1109/AERO.1996.495896

[3] Sung MK, Mudawar I. Correlation of critical heat flux in hybrid jet impingement/micro-channel cooling scheme. International Journal of Heat and Mass Transfer. 2006;**49**: 2663-2672. DOI: 10.1016/j.ijheatmasstransfer.2006.01.008

[4] Rybicki JR, Mudawar I. Single-phase and two-phase cooling characteristics of upward-facing and downward-facing sprays. International Journal of Heat and Mass Transfer. 2006;**49**:5-16. DOI: 10.1016/j.ijheatmasstransfer.2005.07.040

[5] Gao X. Drop impact in spray cooling [thesis]. The University of British Columbia; 2017. DOI: 10.14288/1.0357196

[6] Liang GT, Mudawar I. Review of spray cooling-part 1: Single-phase and nucleate boiling regimes, and critical heat flux. International Journal of Heat and Mass Transfer. 2017;**115**:1174-1205. DOI: 10.1016/j.ijheatmasstransfer.2017.06.029

[7] Shedd TA. Next generation spray cooling: High heat flux management in compact spaces. Heat Transfer Engineering. 2007;**28**:87-92. DOI: 10.1080/01457630601023245

[8] Huddle JJ, Chow LC, Lei S, Marcos A, Rini DP, Lindauer SJ, Bass M, Delfyett PJ. Thermal management of diode laser arrays. In: 16th IEEE Semiconductor Thermal Measurement and Management Symposium; 23 March 2000; San Jose, CA, USA. DOI: 10.1109/STHERM.2000.837078

[9] Yan ZB, Zhao R, Duan F, Wong TN, Toh KC, Choo KF, Chan PK, Chua YS. Spray cooling. In: Ahsam A, editor. Two Phase Flow, Phase Change and Numerical Modeling. Rijeka: Intechopen; 2011. pp. 285-310

[10] Silk EA, Golliher EL, Selvam RP. Spray cooling heat transfer: Technology overview and assessment of future challenges for micro-gravity application. Energy Conversion and Management. 2008;**49**:453-468. DOI: 10.1016/j.enconman.2007.07.046

[11] Tropea C, Roisman IV. Modeling of spray impact on solid surfaces. Atomization and Sprays. 2000;**10**:387-408. DOI: 10.1615/AtomizSpr.v10.i3-5.80

[12] Toda S. A study of mist cooling (first report: Investigation of mist cooling). Heat Transfer: Japanese Research. 1972;**1**:39-50

[13] Toda S. A study of mist cooling (second report: Theory of mist cooling and its fundamental experiments). Heat Transfer: Japanese Research. 1974;**3**:1-44

[14] Chen RH, Chow LC, Navedo JE. Effects of spray characteristics on critical heat flux in subcooled water spray cooling. International Journal of Heat and Mass Transfer. 2002; **45**:4033-4043. DOI: 10.1016/S0017-9310(02)00113-8

[15] Chen RH, Chow LC, Navedo JE. Optimal spray characteristics in water spray cooling. International Journal of Heat and Mass Transfer. 2004;**47**:5095-5099. DOI: 10.1016/j.ijheatmasstransfer.2004.05.033

[16] Mudawar I, Valentine WS. Determination of the local quench curve for spray-cooled metallic surfaces. Journal of Heat Treating. 1989;**7**:107-121

[17] Pautsch AG, Shedd TA. Adiabatic and diabatic measurements of the liquid film thickness during spray cooling with FC-72. International Journal of Heat and Mass Transfer. 2006;**49**:2610-2618. DOI: 10.1016/j.ijheatmasstransfer.2006.01.024

[18] Horacek B, Kiger KT, Kim J. Spray cooling using multiple nozzles: Visualization and wall heat transfer measurements. IEEE Transactions on Device and Materials Reliability. 2004;**4**:614-625. DOI: 10.1109/TDMR.2004.838399

[19] Horacek B, Kiger KT, Kim J. Single nozzle spray cooling heat transfer mechanisms. International Journal of Heat and Mass Transfer. 2005;**48**:1425-1438. DOI: 10.1016/j.ijheatmasstransfer.2004.10.026

[20] Kim J. Spray cooling heat transfer: The state of the art. International Journal of Heat and Fluid Flow. 2007;**28**:753-767. DOI: 10.1016/j.ijheatfluidflow.2006.09.003

[21] Gao X, Li R. Effects of nozzle positioning on single-phase spray cooling. International Journal of Heat and Mass Transfer. 2017;**115**:1247-1257. DOI: 10.1016/j.ijheatmasstransfer.2017.08.095

[22] Mudawar I, Estes K. Optimizing and predicting CHF in spray cooling of a square surface. Journal of Heat Transfer. 1996;**118**:672-679. DOI: 10.1115/1.2822685

[23] Silk EA, Kim J, Kiger K. Spray cooling of enhanced surfaces: Impact of structured surface geometry and spray axis inclination. International Journal of Heat and Mass Transfer. 2006;**49**:4910-4920. DOI: 10.1016/j.ijheatmasstransfer.2006.05.031

[24] Wang YQ, Liu MH, Liu D, Xu K, Chen YL. Experimental study on the effects of spray inclination on water spray cooling performance in non-boiling regime. Experimental Thermal and Fluid Science. 2010;**34**:933-942. DOI: 10.1016/j.expthermflusci.2010.02.010

[25] Visaria M, Mudarwar I. Theoretical and experimental study of the effects of spray inclination on two-phase spray cooling and critical heat flux. International Journal of Heat and Mass Transfer. 2008;**51**:2398-2410. DOI: 10.1016/j.ijheatmasstransfer.2007.08.010

[26] Cheng WL, Liu QN, Zhao R, Fan HL. Experimental investigation of parameters effect on heat transfer of spray cooling. Heat and Mass Transfer. 2010;**46**:911-921. DOI: 10.1007/s00231-010-0631-5

[27] Gao X, Li R. Local heat transfer of spray cooling on a thin film heater. In: 8th International Symposium on Multiphase Flow, Heat Mass Transfer and Energy Conversion; 16-19 December 2016; Chengdu, China

[28] Freund S, Pautsch AG, Shedd TA, Kabelac S. Local heat transfer coefficients in spray cooling systems measured with temperature oscillation IR thermography. International Journal of Heat and Mass Transfer. 2008;**51**:2398-2410. DOI: 10.1016/j.ijheatmasstransfer.2006.09.028

[29] Hsieh S, Tien C. R-134a spray dynamics and impingement cooling in non-boiling regime. International Journal of Heat and Mass Transfer. 2007;**50**:502-512. DOI: 10.1016/j.ijheatmasstransfer.2006.07.023

[30] Karwa N, Kale S, Subbarao PMV. Experimental study of non-boiling heat transfer from a horizontal surface by water sprays. Experimental Thermal and Fluid Science. 2007;**32**:571-579. DOI: 10.1016/j.expthermflusci.2007.06.007

[31] Yang J, Chow LC, Pais MR. Nucleate boiling heat transfer in spray cooling. Journal of Heat Transfer. 1996;**118**:668-671. DOI: 10.1115/1.2822684

[32] Mesler R. A mechanism supported by extensive experimental evidence to explain high heat fluxes observed during nucleate boiling. AICHE Journal. 1976;**22**:246-252. DOI: 10.1002/aic.690220204

[33] Yang J, Pais MR, Chow LC. Critical heat-flux limits in secondary gas atomized liquid spray cooling. Experimental Heat Transfer. 1993;**6**:55-67. DOI: 10.1080/08916159208945369

[34] Estes KA, Mudawar I. Correlation of Sauter mean diameter and critical heat flux for spray cooling of small surfaces. International Journal of Heat and Mass Transfer. 1995;**38**:2985-2996. DOI: 10.1016/0017-9310(95)00046-C

[35] Pais MR, Chow LC, Mahefkey ET. Surface roughness and its effects on the heat transfer mechanism in spray cooling. Journal of Heat Transfer. 1992;**114**:211-219. DOI: 10.1115/1.2911248

[36] Hsieh CC, Yao SC. Evaporative heat transfer characteristics of a water spray on micro-structured silicon surfaces. International Journal of Heat and Mass Transfer. 2006;**49**: 962-974. DOI: 10.1016/j.ijheatmasstransfer.2005.09.013

[37] Zhang Z, Jiang PX, Christopher DM, Liang XG. Experimental investigation of spray cooling on micro-, nano- and hybrid-structured surfaces. International Journal of Heat and Mass Transfer. 2015;**80**:26-37. DOI: 10.1016/j.ijheatmasstransfer.2014.08.085

[38] Chen JN, Xu RN, Zhang Z, Chen X, Ouyang XL, Wang GY, Jiang PX. Phenomenon and mechanism of spray cooling on nanowire arrayed and hybrid micro/nano structured surfaces. Journal of Heat Transfer. 2018;**140**:112401. DOI: 10.1115/1.4039903

[39] Xie JL, Tan YB, Duan F, Ranjith K, Wong TN, Toh KC, Choo KF, Chan PK. Study of heat transfer enhancement for structured surfaces in spray cooling. Applied Thermal Engineering. 2013;**59**:464-472. DOI: 10.1016/j.applthermaleng.2013.05.047

[40] Sodtke C, Stephan P. Spray cooling on micro structured surfaces. International Journal of Heat and Mass Transfer. 2007;**50**:4089-4097. DOI: 10.1016/j.ijheatmasstransfer.2006.12.037

[41] Zhang Z, Jiang PX, Ouyang XL, Chen JN, Christopher DM. Experimental investigation of spray cooling on smooth and micro-structured surfaces. International Journal of Heat and Mass Transfer. 2014;**76**:366-375. DOI: 10.1016/j.ijheatmasstransfer.2014.04.010

[42] Gao X, Li R. Impact of a single drop on a flowing liquid film. Physical Review E. 2015;**92**:053005. DOI: 10.1103/PhysRevE.92.053005

[43] Rioboo R, Tropea C, Marengo M. Outcomes from a drop impact on solid surfaces. Atomization and Sprays. 2001;**11**:155-166. DOI: 10.1615/AtomizSpr.v11. i2.40

[44] Li R, Ashgriz N, Chandra S. Maximum spread of droplet on solid surface: Low Reynolds and Weber numbers. Journal of Fluids Engineering. 2010;**132**:061302. DOI: 10.1115/1.4001695

[45] Clanet C, Beguin C, Richard D, Quere D. Maximal deformation of an impacting drop. Journal of Fluid Mechanics. 2004;**517**:199-208. DOI: 10.1017/S0022112004000904

[46] van Dam DB, Clerc CL. Experimental study of the impact of an ink-jet printed droplet on a solid substrate. Physics of Fluids. 2004;**16**:3403. DOI: 10.1063/1.1773551

[47] Tsai P, Hendrix M, Dijkstra R, Shui L, Lohse D. Microscopic structure influencing macroscopic splash at high Weber number. Soft Matter. 2011;**7**:11325-11333. DOI: 10.1039/C1SM05801K

[48] Chandra S, Avedisian CT. On the collision of a droplet with a solid surface. Proceedings of the Royal Society A. 1991;**432**:13-41. DOI: 10.1098/rspa.1991.0002

[49] Gao X, Li R. Spread and recoiling of liquid droplets impacting solid surfaces. AICHE Journal. 2014;**60**:2683-2691. DOI: 10.1002/aic.14440

[50] Mundo C, Sommerfeld M, Tropea C. Droplet-wall collisions: Experimental studies of the deformation and breakup process. International Journal of Multiphase Flow. 1995;**21**:151-173. DOI: 10.1016/0301-9322(94)00069-V

[51] Xu L. Liquid drop splashing on smooth, rough, and textured surfaces. Physical Review E. 2007;**75**:056316. DOI: 10.1103/PhysRevE.75.056316

[52] Bird JC, Tsai SSH, Stone HA. Inclined to splash: Triggering and inhibiting a splash with tangential velocity. New Journal of Physics. 2009;**11**:063017. DOI: 10.1088/1367-2630/11/6/063017

[53] Vander Wal RL, Berger GM, Mozes SD. The splash/non-splash boundary upon a dry surface and thin fluid film. Experiments in Fluids. 2006;**40**:53-59. DOI: 10.1007/s00348-005-0045-1

[54] Cossali GE, Coghe A, Marengo M. The impact of a single drop on a wetted solid surface. Experiments in Fluids. 1997;**22**:463-472. DOI: 10.1007/s003480050073

[55] Bernardin JD, Stebbins CJ, Mudawar I. Mapping of impact and heat transfer regimes of water drops impinging on a polished surface. International Journal of Heat and Mass Transfer. 1997;**40**:247-267. DOI: 10.1016/0017-9310(96)00119-6

[56] Pasandideh-Fard M, Aziz SD, Chandra S, Mostaghimi J. Cooling effectiveness of a water drop impinging on a hot surface. International Journal of Heat and Fluid Flow. 2001;**22**:201-210. DOI: 10.1016/S0142-727X(00)00086-2

[57] Batzdorf S, Breitenbach J, Schlawitschek C, Roisman IV, Tropea C, Stephan P, Gambaryan-Roisman T. Heat transfer during simultaneous impact of two drops onto a

hot solid substrate. International Journal of Heat and Mass Transfer. 2017;**113**:898-907. DOI: 10.1016/j.ijheatmasstransfer.2017.05.091

[58] Tran T, Staat HJJ, Prosperetti A, Sun C, Lohse D. Drop impact on superheated surfaces. Physical Review Letters. 2012;**108**:036101. DOI: 10.1103/PhysRevLett.108.036101

[59] Staat HJJ, Tran T, Geerdink B, Riboux G, Sun C, Gordillo JM, Lohse D. Phase diagram for droplet impact on superheated surfaces. Journal of Fluid Mechanics. 2015;**779**:R3-R1. DOI: 10.1017/jfm.2015.465

[60] Adera S, Raj R, Enright R, Wang EN. Non-wetting droplets on hot superhydrophilic surfaces. Nature Communications. 2013;**4**:2518. DOI: 10.1038/ncomms3518 (2013)

[61] Jung J, Jeong S, Kim H. Investigation of single-droplet/wall collision heat transfer characteristics using infrared thermometry. International Journal of Heat and Mass Transfer. 2016;**92**:774-783. DOI: 10.1016/j.ijheatmasstransfer.2015.09.050

[62] Yarin AL, Weiss DA. Impact of drops on solid surfaces: Self-similar capillary waves, and splashing as a new type of kinematic discontinuity. Journal of Fluid Mechanics. 1995;**283**:141-173. DOI: 10.1017/S0022112095002266

[63] Roisman IV, Tropea C. Impact of a drop onto a wetted wall: Description of crown formation and propagation. Journal of Fluid Mechanics. 2002;**472**:373-397. DOI: 10.1017/S0022112002002434

[64] Cossali GE, Marengo M, Coghe A, Zhdanov S. The role of time in single drop splash on thin film. Experiments in Fluids. 2004;**36**:888-900. DOI: 10.1007/s00348-003-0772-0

[65] Rioboo R, Bauthier C, Conti J, Voue M, De Connick J. Experimental investigation of splash and crown formation during single drop impact upon wetted surfaces. Experiments in Fluids. 2003;**32**:648-652. DOI: 10.1007/s00348-003-0719-5

[66] Alghoul SK, Eastwick CN, Hann DB. Normal droplet impact on horizontal moving films: An investigation of impact behaviour and regimes. Experiments in Fluids. 2011;**50**:1305-1316. DOI: 10.1007/s00348-010-0991-0

[67] Che ZZ, Deygas A, Matar OK. Impact of droplets on inclined flowing liquid films. Physical Review E. 2015;**92**:023032. DOI: 10.1103/PhysRevE.92.023032

[68] Gao X, Li R. Cooling enhancement of drop impact on flowing liquid film. In: The 2nd Thermal and Fluids Engineering Conference and 4th International Workshop on Heat Transfer; 2-5 April 2017; Las Vegas, Nevada, USA. DOI: 10.1615/TFEC2017.hte.018061

[69] Gao X, Kong LJ, Li R, Han JT. Heat transfer of single drop impact on a film flow cooling a hot surface. International Journal of Heat and Mass Transfer. 2017;**108**:1068-1077. DOI: 10.1016/j.ijheatmasstransfer.2016.12.106

[70] Qiu L, Dubey S, Choo FH, Duan F. Splashing of high speed droplet train impinging on a hot surface. Applied Physics Letters. 2015;**107**:164102. DOI: 10.1063/1.4934531

[71] Qiu L, Dubey S, Choo FH, Duan F. The transitions of time-independent spreading diameter and splashing angle when a droplet train impinging onto a hot surface. RSC Advances. 2016;**6**:13644. DOI: 10.1039/C5RA26314J

[72] Soriano GE, Zhang TL, Alvarado JL. Study of the effects of single and multiple periodic droplet impingements on liquid film heat transfer. International Journal of Heat and Mass Transfer. 2014;**77**:449-463. DOI: 10.1016/j.ijheatmasstransfer.2014.04.075

[73] Zhang TL, Alvarado JL, Muthusamy JP, Kanjirakat A, Sadr R. Numerical and experimental investigations of crown propagation dynamics induced by droplet train impingement. International Journal of Heat and Fluid Flow. 2016;**57**:24-33. DOI: 10.1016/j.ijheatfluidflow.2015.10.003

[74] Zhang TL, Alvarado JL, Muthusamy JP, Kanjirakat A, Sadr R. Heat transfer characteristics of double, triple and hexagonally-arranged droplet train impingement arrays. International Journal of Heat and Mass Transfer. 2017;**110**:562-575. DOI: 10.1016/j.ijheatmasstransfer.2017.03.009

[75] Gao X, Li R. Impact of a drop burst flow on a film flow cooling a hot surface. International Journal of Heat and Mass Transfer. 2018;**126**:1193-1205. DOI: 10.1016/j.ijheatmasstransfer.2018.06.042

[76] Chen H, Cheng WL, Peng YH, Zhang WW, Jiang LJ. Experimental study on optimal spray parameters of piezoelectric atomizer based spray cooling. International Journal of Heat and Mass Transfer. 2016;**103**:57-65. DOI: 10.1016/j.ijheatmasstransfer.2016.07.037

[77] Gao X, Li R. Heat transfer of drop burst impact on a flowing film. In: 16th International Heat Transfer Conference; 10-15 August 2018; Beijing, China

Chapter 4

Heat Pump-Based Novel Energy System for High-Power LED Lamp Cooling and Waste Heat Recovery

Jiwen Cen, Zhibin Li, Yiwei Wang, Fangming Jiang,
Shaoxiong Liao and Fuwen Liang

Additional information is available at the end of the chapter

http://dx.doi.org/10.5772/intechopen.78322

Abstract

Unlike incandescent light bulb, which radiates heat into the surroundings by infrared rays, light emitting diode (LED) traps heat inside the lamp. This fact increases the difficulty of cooling LED lamps, while it facilitates the recovery of the generated heat. We propose a novel energy system that merges high-power LED lamp cooling with the heat pump use; the heat pump can cool the LED lamp and at the same time recover the waste heat. In this way, a high percentage of the energy consumed by the LED lamp can be utilized. In this work, we developed a prototype of this energy system and conducted a series of experimental studies to determine the effect of several parameters, such as cooling water flow rate and LED power, on the LED leadframe temperature, compressor power consumption, and system performance. The experimental results clearly indicate that the energy system can lead to substantial energy savings.

Keywords: LED, heat pump, heat recovery, cooling, energy saving

1. Introduction

Light emitting diode (LED) is a promising solid state light source due to its high energy efficiency, eco-friendliness, small volume, long lifetime, quick response, low driving voltage, and good structural integrity. Compared with traditional incandescent light bulbs, an LED lamp only needs 10–20% of electric power to produce the same luminous flux [1].

LED is a cold light source, in which a photon is emitted when an electron transits from a high energy band to a low one inside LED. The wavelength of the emitted light is determined by the energy gap of the light emitting semiconductor material; it is usually fixed at a certain visible light wavelength, which does not include the infrared range, while the light of the incandescent light bulb includes a wide range of wavelengths including visible light and infrared rays. Therefore, the heat generated inside LEDs cannot be radiated outside by infrared radiation as it occurs with the incandescent light bulb. So far, the luminous efficiency of LED is only about 20–30%. The remainder of the power input, about 70–80% in total, is eventually converted into heat. If the heat is not dissipated timely, the LED temperature will quickly rise. Once the pn junction temperature reaches its maximum allowed temperature, which is typically around 150°C, LED will burn out. In industry experience, LED will work efficiently when its leadframe temperature is below 60–70°C.

The major disadvantage of LED is its inability of radiating the generated heat; therefore, effective cooling measures should be taken to guarantee the expected life time of the LED. In the open literature, different cooling approaches to improve LED lamp performance have been reported. In the marketplace, traditional LED lamps dissipate heat into ambient air [2–17]. For example, Dong et al. [2] performed numerical simulations to determine the influence of heat sink design on LED lamp thermal dissipation; their specific aim was to optimize the number of fins and thickness of the heat sink. They found that these two parameters were key parameters for the shape and mass of the heat sink in practical applications. Tang et al. [3] developed a novel columnar heat pipe leadframe for a high-power LED device. They demonstrated that the luminous efficacy of the LED device with the columnar heat pipe leadframe is 19.2% higher than that with the conventional copper leadframe. Deng et al. [4] proposed an active cooling solution using liquid metal as the coolant for high-power LED lamps. They found that liquid metal cooling is a powerful way to dissipate heat from high-power LED lamps, and the fabrication of practical liquid metal cooling devices is feasible and their use is promising. Lin et al. [5] conducted an experimental study to investigate the heat transfer characteristics of an aluminum plate oscillating heat pipe (OHP), which consisted of parallel and square channel arrangements. Their experimental results indicate that the temperature of the LED decreases significantly while being cooled by natural convection when a plate OHP was used in the LED heat sink. Dong et al. [6] developed a heat sink with ionic wind for LED cooling. Their experiments and computational fluid dynamics (CFD) analysis confirmed the advantages of the heat sink with ionic wind for LED cooling. Wang et al. [7] also suggested using a needle-to-net type ionic wind generator based on corona discharge for high-power LED cooling. Their experimental results indicate that the designed ionic wind generator had good cooling performance close to cooling fan, coupled with lower energy consumption and less mechanically induced noise. Yung et al. [8] studied the thermal performance of a high-brightness LED array on printed circuit board (PCB) under different placement configurations. They proposed a new LED placement configuration to lower the individual LED temperature in the array by 26°C. Park et al. [9] developed an LED downlight consisting of a chimney and a radial heat sink. They concluded that installing the chimney can improve the heat sink cooling efficiency by up to 20% as compared to the installation of the hollow cylinder heat sink. The same team investigated an inclined cross-cut cylindrical heat sink in an attempt to improve the energy conversion and management of LED light bulbs [10]. They showed that when the fins had an angle of inclination of 25–30°, the thermal resistance was the lowest. However, the cooling performance decreases when the angle of inclination is greater than 50°. Xiao et al. [11] developed an automatic cooling device for thermal management of high-power LEDs. The device consists of a microcontroller, heat pipes, and a speed adjustable fan. The thermal resistances R_{sa} (from the heat sink to the ambient) and R_{ja} (from the LED chip to the ambient) of the cooling system operating at 12 Ware 0.37 and 5.95°C/W, respectively. The total power consumption of the cooling system is less than 1.58 W. Sömen et al. [12] studied by using ANSYS CFX 14 software the cooling effects of fin design, fin material, and free and forced convection on the junction temperature of automotive LED headlights. They presented a new methodology for defining the optimum cylindrical fin structure, and they also indicated that as the ambient temperature increases, especially in high-power LED lights, proper fin usage becomes essential. Jeng et al. [13] systematically studied the heat transfer characteristics of a porous green building material and the enhancement of LED lamp heat sink when LED lamp is inserted into this porous material. They demonstrated that the closed-cell aluminum-foam ceiling did help in the cooling of LED lamp. Wu et al. [14] designed a phase-change material (PCM) base heat pipe heat sink (PCM-HP heat sink) that consists of a PCM base, adapter plate, heat pipe, and finned radiator. The results show that the heat sink possesses remarkable thermal performance owing to the reduction of the LED heating rate and peak temperature and an excellent antithermal-shock capacity. Sufian and Abdullah [15] reported the enhancement of the heat transfer in high-power LEDs by a combination of piezoelectric fans and a heat sink. The results showed that the dual fans enhanced the heat transfer performance by approximately 3.2 times, while the quadruple fans enhanced heat sink of the LEDs by 3.8 times compared to natural convection. Zhao et al. [16] presented a study of the thermal performance of conventional plate-fin heat sinks and novel cooling device integrated with heat conductive plates (HCPS) for the application in high-power LED headlight. The results showed that the thermal performance of the heat sink with HCPS is better than the one only with the heat sink cooling system. Kang et al. [17] investigated a new cooling method of ethanol direct-contact phase-change immersion cooling in the thermal management of high-power LED. The experimental results showed that the ethanol phase-change immersion cooling is an effective way to make sure high-power LED work reliably and high efficiently.

In all the above mentioned approaches used for LED cooling, heat is simply dissipated with no further usage. Taking into consideration the influence of this heat dissipation on the surroundings, Ahn et al. [18, 19] proposed a new methodology to integrate LED lighting with heating, ventilation, and air conditioning systems to prevent this heating buildup in buildings. The heat sinks of LED lamps are placed inside the ventilation ducts; therefore, most of the LED heat is removed to outdoors by the duct air flow, avoiding in this way a further increase of the building thermal load.

An advantage for LED is related to being a cold light source, which offers the potential of heat recovery; as already mentioned, unlike incandescent light bulbs, which transfer heat into the surroundings by infrared radiation, LED traps heat inside the lamp if no efficient heat dissipation device to help it, which facilitates its recovery. However, the heat removal and recovery should proceed in such a way that maximum allowable temperature is not reached. If the aim is to use this heat for higher temperature applications, then the heat pump is the appropriate device.

In this paper, an integrated system combining a high-power LED lamp with a heat pump is proposed to simultaneously cool down the LED lamp and provide hot water using the LED waste heat. This approach allows that most of the electric energy consumed by the LED be utilized, which greatly improves the energy efficiency. This combined system can be used to reduce building thermal load as indicated in the work by Ahn et al. [18, 19]. In addition, it is particularly suited to the scenarios that both high-power light and hot water are needed. For example, in railway station halls, airports, theaters, and sport stadiums, not only sufficient illumination is needed, but also a great amount of warm water is needed for sanitary.

2. Experimental prototype

Schematic of the experimental setup is shown in **Figure 1**, and the photo of the actual experimental setup is shown in **Figure 2**. The experimental prototype of the integrated system mimics a high pole LED lamp.

This system consists of an LED lamp acting as an evaporator for a heat pump, a regulator for oil returning, a compressor, a condenser, an expansion valve, and some connecting pipes. The structure of the lamp (or evaporator) is a hermetic hemisphere of 420 mm in diameter, which is made of stainless steel. The "opening" of the hemisphere is facing upward, and six equidistant round openings with a diameter of 38 mm are made on hemisphere surface with their centers at a vertical distance of 160 mm from the top plane. These round opening apertures are capped by the LED modules, which are tightly sealed by flanges and seal rings to keep the system hermetically sealed. In this arrangement, heat from the LED modules is transferred by pool boiling of the refrigerant inside the evaporator.

For experimental convenience with different LED power values, the LED modules are replaced with six aluminum blocks with inserted cartridge heaters; in this way, the heat inputs can be

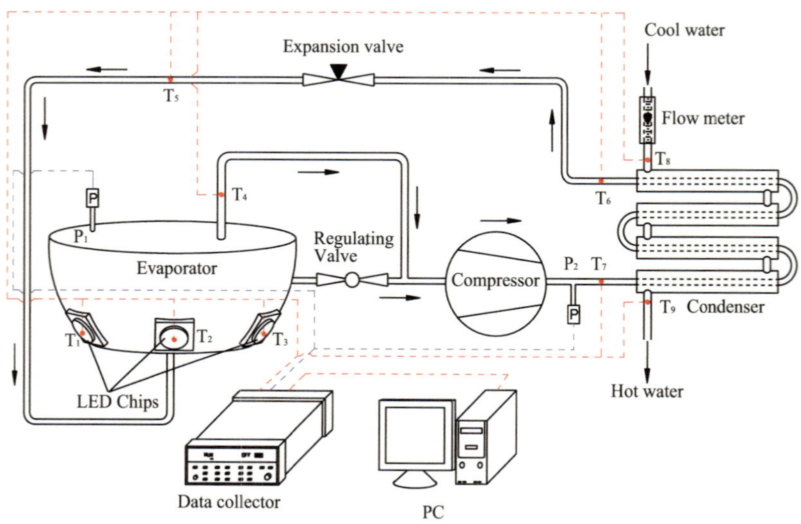

Figure 1. Schematic of the experimental setup. Dashed lines indicate the data acquisition connections; T denotes the temperature measuring point and P the pressure monitoring point.

Figure 2. Photo of the experimental prototype.

easily controlled and measured. All surfaces of the system components, including the lamp and pipes, are wrapped with plastic insulation foam for thermal insulation.

Model number of the compressor is Hitachi SL242CV-C7LU with a rated power of 1150 W. It is an invariable frequency compressor. Further system improvement should be considered to use a variable frequency compressor with an inverter. The condenser is a double-pipe coaxial heat exchanger produced by Hangzhou Shenshi Energy Conversion Technology Co., Ltd. Its model number is QH2P286254. The diameter of the outer tube is 38 mm and inner tube 22 mm. Total heat exchange surface of the condenser is about 1 m^2, and water flows through the inner tube of the coaxial heat exchanger and R22 refrigerant flows through the annular region. A needle valve serves as the expansion valve of the heat pump. All the heat pump components are connected with 8 mm diameter pipes. Here, it has to be point out that R22 is going to be phased down and other new environmental friendly refrigerants, such as R410A, R290, R134A, R407c, R717, CO_2, etc., should be adopted. But which refrigerant is the most suitable for the heat-pump-based LED heat recovery system is further needed to be investigated. In this paper, R22 is used to build a base line system for future new refrigerant comparison because R22 is one of the most commonly used refrigerants in the past several decades.

The temperature at several points of the system is measured with K-type thermocouples with an accuracy of ±0.3°C. High-side and low-side pressure of the heat pump system are measured with pressure transducers model MB300,with a measuring range of 0–5 MPa with an uncertainty of ±0.2% full scale (FS), produced by Nanjing GOVA technology Co., Ltd. Temperature and pressure signals are collected by an Agilent data acquisition system and then transferred to a computer for further data processing. The water flow rate in the condenser is measured using a rotameter with an uncertainty of ±1% FS, in the range of 0–0.4 l/s. The compressor power consumption is measured by a digital power meter in a range of 0–12 kW, with an uncertainty of ±1% FS. The calculated relative uncertainty range of coefficient of performance (COP) is approximately ±5%. Measurement points T_1–T_3 are LED leadframe temperature, T_4 is evaporator outlet temperature, T_5 is evaporator inlet temperature, T_6 is condenser outlet temperature, T_7 is condenser inlet temperature, T_8 is water inlet temperature, T_9 is water outlet temperature, P_1 is evaporator pressure, and P_2 is condenser pressure.

The working principle of the integrated system is as follows: refrigerant with quality close to zero flows through the expansion valve and enters the LED lamp, which serves as the evaporator for the heat pump. In the evaporator, the refrigerant evaporates, while it absorbs heat from the LEDs. The LED lamp is kept at the appropriate temperature by regulating the expansion valve. The refrigerant in vapor phase leaves the evaporator and enters the low pressure side of the compressor; at the exit of the compressor, the refrigerant is in the superheated state at relatively high temperature and pressure. The refrigerant then passes through the condenser, where it releases heat to the coolant (water). The cooled down refrigerant leaving the condenser enters the expansion valve, completing in this way a cycle. In reality application, a hot water tank should be used in case that hot water demand (e.g., domestic hot water needs) may be consistent with LED cooling demand.

In this work, as already mentioned, the LED lamp in the experimental setup is designed for a high pole LED lamp, a type of lamp typically encountered in large squares, harbors, and airports. In the present experimental setup, the quantity of refrigerant charged into the system should be sufficient to have the back sides of the LED modules fully immersed in the liquid refrigerant, when the system is operating at steady state conditions. In this way, effective heat transfer from the modules to the liquid refrigerant is guaranteed. The outlet of the evaporator is on the top side of the lamp (evaporator), where the refrigerant in vapor phase accumulates. However, during the operation compressor, oil contaminates the refrigerant in the liquid phase by floating on it and/or by mixing with it. To partially overcome this problem, an oil outlet was located approximately close to the level of the liquid refrigerant surface inside the evaporator; the flow rate of the oil coming through the outlet is controlled by a regulating valve. There are two modes to control the valve. The first is to open the valve constant at an appropriate small opening. The second is to open the valve intermittently. The primary aim of this valve is to prevent that too much liquid refrigerant flows out of the evaporator and two much oil accumulate inside the evaporator.

3. System performance evaluation

In the integrated system of LED lamp and heat pump, the LED leadframe temperature and the heat pump COP are the parameters of primary concern in the performance evaluation.

The COP of heat pump is defined as,

$$COP = \frac{Q_{water}}{W_{comp}} \tag{1}$$

where Q_{water} refers to the heat removed by the water flowing through the condenser, and W_{comp} refers to the power consumption of the compressor. Consequently, to conduct a parametric analysis, we investigate the effect of heating power and water flow rate on the performance of the integrated system. The heating power varies in the range of 1400–2400 W, which corresponds to the heat generation of LED lamps rated at approximately 2000–3400 W power, which are typical values for high-pole LED lamps. The water flow rate passing through the condenser is in the range of 0.020–0.240 kg/s; these flows are sufficiently small to produce relatively high-temperature water, which can be used as domestic hot water, while maintaining the temperature of the LED lamp with relatively minor variations.

The heat transferred to the water, Q_{water}, is calculated by using the following relation:

$$Q_{water} = c_w \dot{m}_w (T_{w\_out} - T_{w\_in}) \tag{2}$$

where c_w refers to the water specific heat, $\dot{m}_w$ refers to the water mass flow rate, T_{w_out} refers to the water outlet temperature, and T_{w_in} refers to the water inlet temperature.

Based on the energy balance for the compressor with the assumption the process is adiabatic, the refrigerant mass flow rate, $\dot{m}_r$, is determined as follows:

$$\dot{m}_r = \frac{W_{comp}}{h(T_{c\_in}, P_c) - h(T_{e\_out}, P_e)} \tag{3}$$

where T_{c_in} refers to the refrigerant temperature at the condenser inlet, T_{e_out} refers to the refrigerant temperature at the evaporator outlet, P_c and P_e refer to condenser and evaporator pressure, respectively, and h refers to refrigerant specific enthalpy determined for the specified state (T,P).

Compressor specific power consumption SW_{comp} is defined as

$$SW_{comp} = \frac{W_{comp}}{\dot{m}_r} \tag{4}$$

The influence of the water flow rates on the LED leadframe temperature is reported in **Figure 3**. By increasing the water flow rate, the temperature of the LED leadframe experiences a slight decrease. This observation is consistent with the heat rate of the condenser, which is constrained by the refrigerant side heat transfer coefficient; therefore, although higher water flow rates yield higher water side heat transfer coefficients, the impact on the overall heat transfer coefficient is of no great significance. Consequently, the temperature of the LED leadframe varies accordingly. This characteristic of the integrated system should be seen as an advantage, considering that a severe escalation of LED temperature will not occur when it is required high-temperature water at reduced flow rate.

Figure 4 reports the water temperature at the condenser outlet as a function of mass flow rate and heating power. As expected, with the increase of water flow rate, the outlet temperature decreases, and eventually with further increase of the flow rate, it would tend to the value of the water inlet temperature. From this figure, it can be observed that at the lower end of the

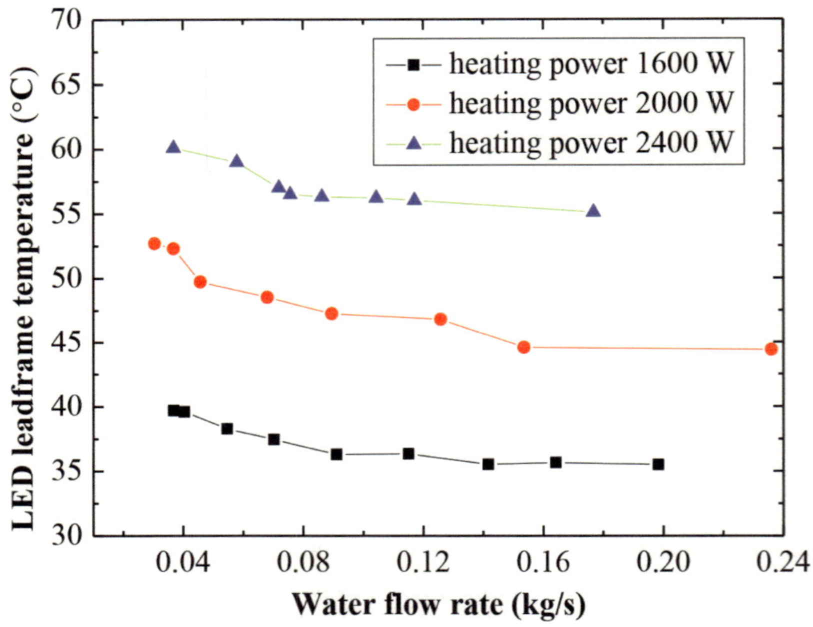

Figure 3. Temperature of the LED leadframe for different rates of water mass flow and heating power.

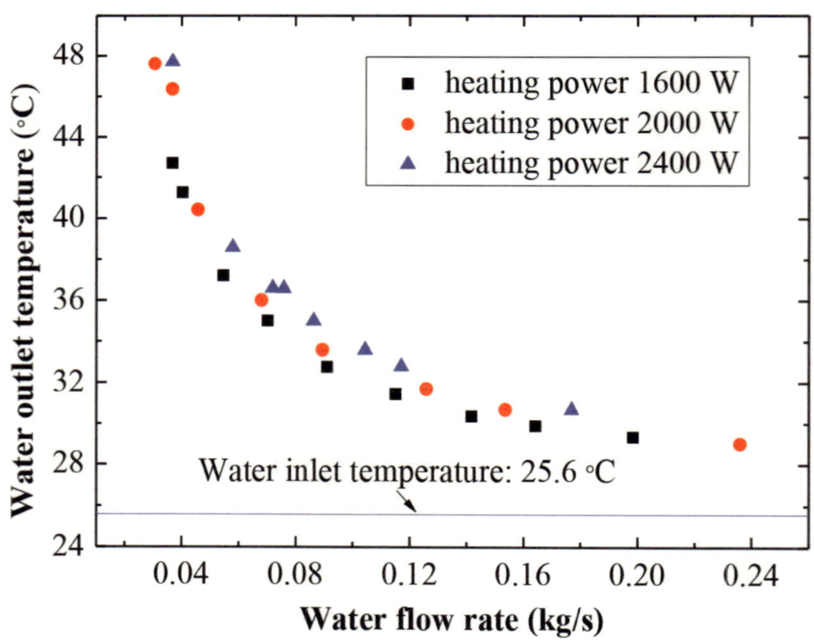

Figure 4. Water outlet temperature as a function of the water flow rate and heating power.

mass flow rates tested, the water outlet temperature can reach temperatures of 48°C; not surprisingly, for the tested higher water flow rates, the increase in the temperature from the inlet to the outlet is practically insignificant.

Figure 5 shows the influence of the condenser water flow rate on the compressor power consumption for different values of the heating power. In this figure, it can be observed that the increase of water flow rate reduces power consumption of the compressor, which is an interesting result. Its rationale is related to the fact that an increase of the water flow rate yields an increase in the condenser heat transfer rate, in this way, setting the temperature of the superheated vapor leaving the compressor. Therefore, the power consumed by the compressor is directly related to its outlet temperature as well established by thermodynamic principles. However, consistent with the results presented in **Figure 3**, further increase of the water mass flow rate toward the upper end of the tested range leads to no significant reduction of compressor power consumption.

Taking into consideration the results, in particular those reported in **Figures 3** and **5**, respectively, they indicate that increasing values of mass flow rate yield higher values of COP, as it will be shown in **Figure 6**. However, higher values of the water flow rate may lead to outlet temperatures too low for practical usage, which is illustrated by the results presented in **Figure 4**.

Figure 6, as already mentioned, reports the influence of water flow rate on the heat pump COP, it can be noted that the system COP increases with increasing water flow rate. COP experiences a sharp increase from 2.4 to about 3.56 when the water flow rate increases from 0.03 to 0.18 kg/s, respectively, under the heating power of 2400 W. This result is justified on the

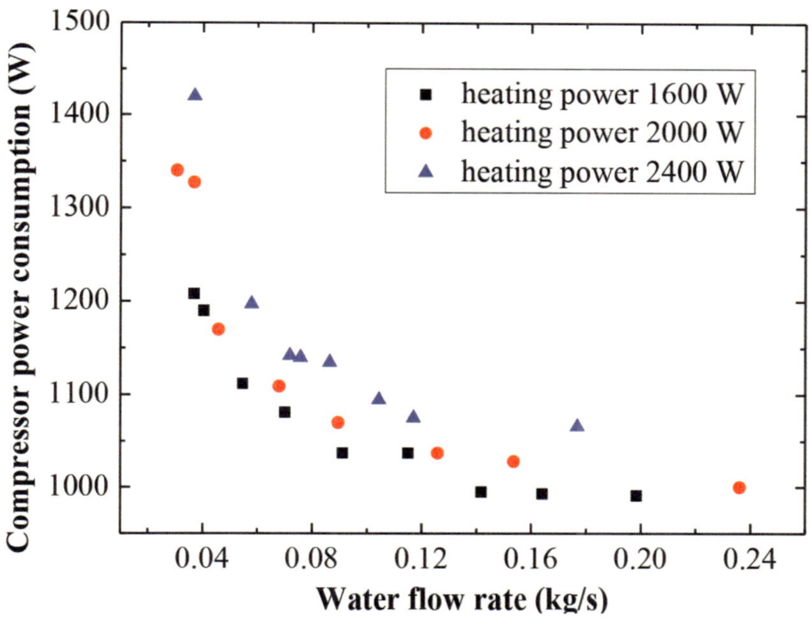

Figure 5. Compressor power consumption as a function of the water flow rate and heating power.

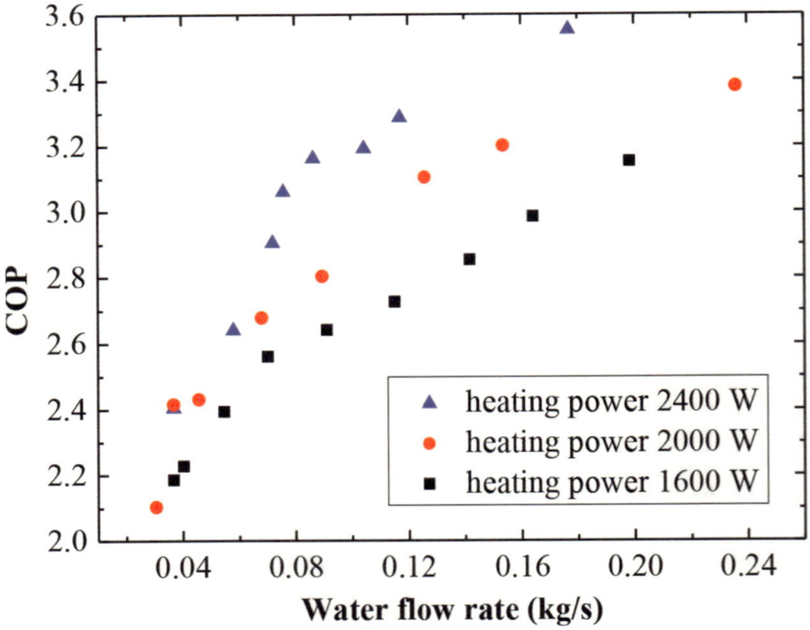

Figure 6. The effect of water flow rate and heating power on the system COP.

basis of data already reported in particular that in **Figures 3** and **5**. In addition, the higher the heating power, the higher the system COP is. It is because that higher heating power will result in higher evaporation temperature and higher evaporation pressure, which reduces the pressure ratio of the compressor and consequently improves the system COP.

The data presented in **Figure 3** are further analyzed in **Figure 7**, and it is clearly shown that the LED leadframe temperature increases with increasing heating power. The experimental results indicate that for a 2400 W heating power level, the LED leadframe temperature can be kept within 60°C. For lower values of LED power, the LED leadframe temperature can even be lower than the surrounding air temperature. Consequently, taking into consideration the correlation between LED lifetime and its junction temperature, the proposed integrated system has the potential of prolonging the lifetime of the LEDs. However, it should be mentioned that in practice, the LED temperature cannot be set lower than that of the dew point of the atmosphere; otherwise, the condensation of the water vapor in the air will damage the electronic components of the LED. Moreover, as already stated, the LEDs should be set at an appropriate temperature, which may lead to an operation close to the maximum energy efficiency of the integrated system.

Also **Figure 8**, further to rearrange data presented in **Figure 5**, indicates that the power absorbed by the compressor increases significantly with increasing LED power; as before, this can be explained on the basis of the increasing flow rate of refrigerant through the compressor, when the heating power increases. The increased mass flow rate, assuming the inlet and outlet states remain the same, will lead to increased power consumption.

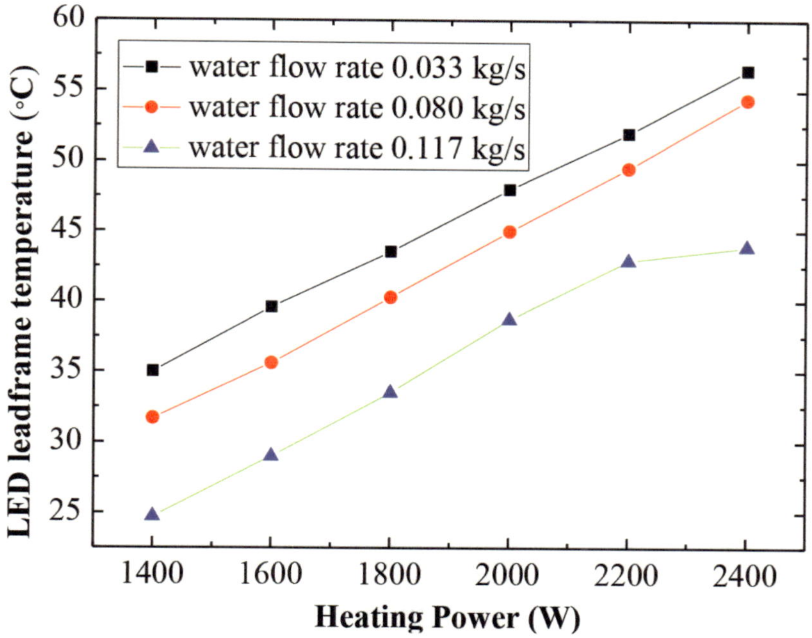

Figure 7. LED leadframe temperature as a function of heating power for different water flow rates.

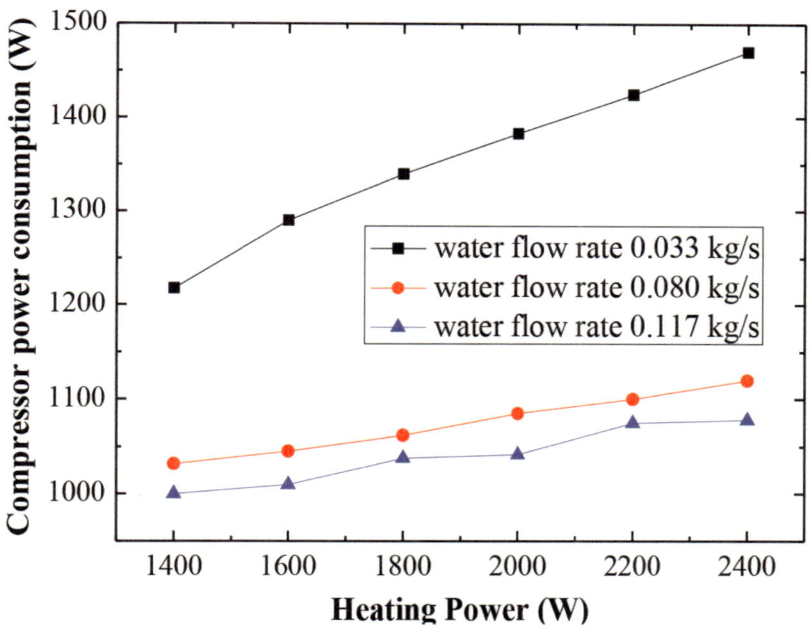

Figure 8. Compressor energy consumption as a function of heating power for different water flow rates.

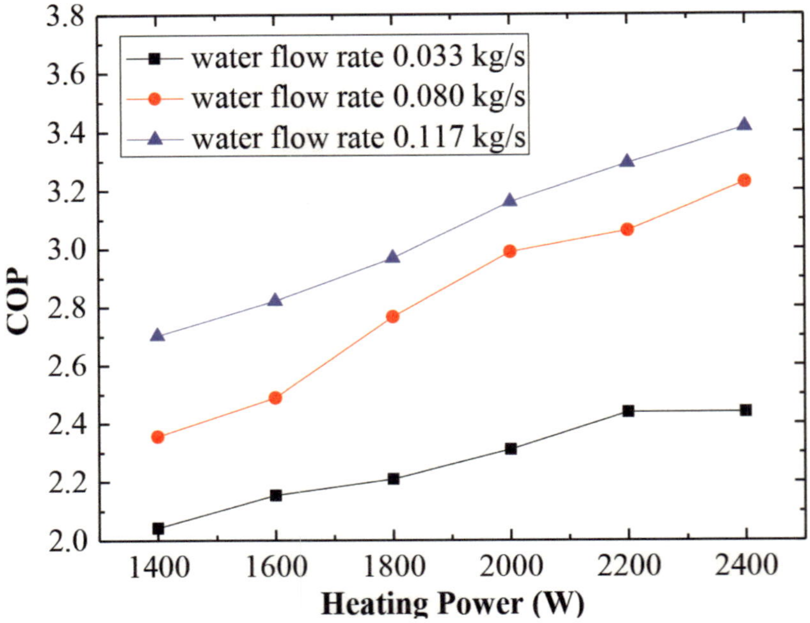

Figure 9. Evolution of system COP with various heating powers.

Figure 9 shows the influence of the total heating power on the system COP. The system COP increases nearly linearly with the total heating power. The increase of the total heating power yields an increase in the evaporating temperature; therefore, the temperature difference between average high temperature T_H side and low temperature T_L side of the heat pump system is reduced. In these conditions considering the COP of the ideal Carnot heat pump cycle, $COP_{ideal} = T_H/(T_H - T_L)$, the heat pump COP of the system should increase. However, in this particular case, the evaporating temperature is limited by the maximum allowed temperature for the LED pn junction.

Table 1 lists the experimental data acquired for the present work when the water inlet temperature is 25.6°C and the total heating power is 2000 W; it encompasses, for different water mass flow rates, the refrigerant inlet and outlet temperatures of evaporator and condenser, respectively, and their refrigerant pressures, water outlet temperature, and compressor power consumption. In addition, the calculated quantities—heat transferred to the water, refrigerant mass flow rate, and specific power consumption of the compressor—are also included in **Table 1**.

As already discussed, for a specific heating power, the water mass flow rate drives the system, as it sets the temperature at the outlet of the compressor (or condenser inlet temperature). Therefore, by increasing the water flow rate, the condenser and evaporator pressures decrease as well as the condenser outlet temperature and the evaporator inlet and outlet temperatures. It is interesting to note that the decrease of the water mass flow rate leads to, although minor, decrease of the refrigerant mass flow rate; consequently, the specific power consumption

$\dot{m}_w \times 10^{-3}$ (kg/s)	T_{e_in} (°C)	T_{e_out} (°C)	T_{c_in} (°C)	T_{c_out} (°C)	$P_e \times 10^6$ (Pa)	$P_c \times 10^6$ (Pa)	T_{w_out} (°C)	W_{comp} (W)	Q_{water} (W)	$\dot{m}_r \times 10^{-3}$ (kg/s)	$SW_{comp} \times 10^3$ (W/(kg/s))
236	−3.1	5.2	39.4	29.1	0.41	1.15	29.01	1000	3379	17.94	55.75
154	−3.0	5.4	40.0	29.2	0.42	1.19	30.7	1028	3289	17.54	58.62
90	−1.6	6.5	40.8	31.5	0.44	1.26	34.08	1070	3187	17.33	61.76
68	−0.2	8.1	42.4	33.7	0.47	1.35	36.5	1109	3113	17.21	64.44
46	2.6	10.1	45.7	37.4	0.51	1.47	40.62	1170	2885	16.32	71.69
31	6.3	13.8	56.1	43.8	0.59	1.75	47.6	1340	2818	16.30	82.22

$\dot{m}_w$: water mass flow rate, T_{e_in}: evaporator inlet temperature, T_{e_out}: evaporator outlet temperature, T_{c_in}: condenser inlet temperature, T_{c_out}: condenser outlet temperature, P_e: evaporator pressure, P_c: condenser pressure, T_{w_out}: water outlet temperature, W_{comp}: compressor power consumption, Q_{water}: condenser heat transfer rate, $\dot{m}_r$: refrigerant mass flow rate, and SW_{comp}: specific power consumption of the compressor.

Table 1. Inlet and outlet temperature of evaporator and condenser and their inside pressures with various water flow rates when total heating power is 2000 W and water inlet temperature is 25.6°C.

experiences a significant increase, which indicates that the difference between the outlet and inlet temperatures of the compressor increases. The heat transfer rate in the condenser, considering that it is an annular heat exchanger, is dominated by the refrigerant side heat transfer coefficient; notwithstanding, the fact that the water side heat transfer coefficient will increase with increasing water mass flow rate.

For the system, as a whole, for a specific heating power generated by the aluminum blocks with inserted cartridge heaters, which mimic LEDs, the increase of the water mass flow rate will have the effect of reducing the temperature in the evaporator (lamp) and consequently the corresponding saturation pressure, resulting in this way an increase in boiling activity yielding an increase in the refrigerant mass flow rate. As expected, the decrease in evaporator pressure tends to be less than that at the condenser. Therefore, by increasing the water mass flow rate, the compressor pressure ratio decreases and, consequently, the COP also increases.

4. The thermodynamic performance limit of the system

According to the first law of thermodynamics and omitting heat leakage from the system to the surroundings, the energy conservation equation of the system can be written as

$$Q_{water} = W_{comp} + (1 - \eta) W_{\text{LED}} \tag{5}$$

where η refers to the LED luminous efficiency, and W_{LED} refers to the power consumed by the LED chips.

Theoretically, the heat pump Carnot efficiency is $COP_{\text{ideal}} = T_H/(T_H - T_L)$, then considering the definition of the system COP, we have

$$COP = f\frac{T_H}{(T_H - T_L)} = 1 + \frac{(1-\eta)W_{LED}}{W_{comp}} \tag{6}$$

where f is a ratio of the real COP to COP_{ideal}. By increasing the LED power, T_L will also increase to keep the energy balance. However, the combined increase of refrigerant temperature in the evaporator and LED power results in an even higher LED chip temperature to achieve the required heat transfer between the LED chip and the refrigerant. Irreversibility due to the heat transfer, compression, and throttling processes will keep the values of the heat pump COP much lower than those for the Carnot heat pump. In addition, the increase of the evaporator temperature will result in higher compressor exhausting temperature, which may not satisfy the rated operating condition of the compressor, and its efficiency will deteriorate. In these conditions, Q_{water} will not increase as much as the increase of LED power and the additional heat will be accumulated in the system, in particular in the evaporator. In these circumstances, when the LED power reaches a certain quantity, there is a possibility that the refrigerant liquid phase will be reduced to the point that the LED chip reaches a temperature that will cause burnout.

As shown in **Figure 10**, when the heating power is 2400 W, the LED leadframe temperature can be kept at a stable temperature of 55°C. However, when heating power is increased to 2500 W, the LED leadframe temperature shoots up after a period time of heat accumulation. In the experiment, we also used another compressor with a rated power of 953 W to test its working limit. Its highest allowed heating power was about 1800 W, both compressors have almost the same COP with values higher than 3, when they operate at their highest allowed heating powers.

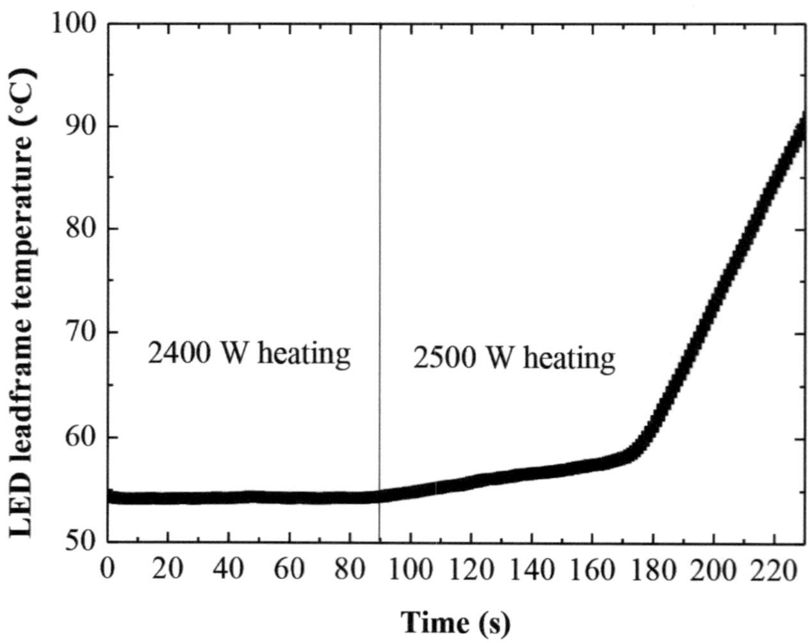

Figure 10. Working limit of the system.

5. Conclusions

The present work proposes an integrated system for temperature control and heat recovery to operate in conjunction with an LED lamp. The integrated system consists of a heat pump in which the lamp itself operates as the heat pump evaporator. The temperature control of the LED can be achieved through the coolant (water) flow rate through the condenser. To demonstrate the concept, it was developed as an experimental setup, in which the evaporator mimics a high pole LED lamp. After extensively testing the apparatus, the most significant findings are as follows:

1. Water flow rate has relatively minor influence on the LED leadframe temperature; therefore, when higher temperature water is required, the water flow rate can be reduced without having an escalation of the LED temperature.

2. The increase of water flow rate yields a decrease of the compressor power consumption; however, further increase of the flow does not lead to significant reduction of compressor power consumption.

3. The increase of water flow rate causes the COP to increase, and further increase of the flow only results in minor augmentation of COP.

4. For the highest total heating power tested, i.e., 2400 W, which is equivalent to an LED lamp with a total power of ~3000 W, the LED leadframe in this experimental setup can be kept at 60°C level; for lower values of LED power, the LED leadframe can even be lower than the surrounding air temperature.

5. LED power increase results in sharp increase of compressor energy consumption.

6. System COP presents a nearly linear relation with LED power.

7. The integrated system working limit occurs when LED power reaches a quantity, which causes the temperature of the compressor to exceed its maximum specified value; under these circumstances, the LED chip may reach burnout conditions.

In conclusion, this integrated system adopts an active method to simultaneously achieve waste heat recovery and LEDs cooling down. For an appropriate compressor rated power, LEDs will be kept within optimal temperature range. In addition, the recovered heat will come in the form of hot water, which may find multiple applications. Moreover, the integrated system avoids dissipating heat to the surroundings, minimizing in this way the environmental impact, and as it does not require a large aluminum finned heat sinks, there is an obvious gain in terms of compactness.

Acknowledgements

This work was supported by Science and Technology Program of Guangzhou, China (grant no. 201604010018), and Science and Technology Program of Guangdong, China (grant no. 2016A010104010). The authors also thank the CAS President's International Fellowship

Initiative Visiting Fellow, Prof. Antonio CM Sousa, for helpful and constructive suggestions when preparing the paper.

Author details

Jiwen Cen[1], Zhibin Li[1], Yiwei Wang[1], Fangming Jiang[1]*, Shaoxiong Liao[2] and Fuwen Liang[2]

*Address all correspondence to: fm_jiang2000@yahoo.com

1 Laboratory of Advanced Energy Systems, CAS Key Laboratory of Renewable Energy, Guangdong Key Laboratory of New and Renewable Energy Research and Development, Guangzhou Institute of Energy Conversion, Chinese Academy of Sciences (CAS), China

2 Tiger Fire (Guangzhou) Technology Co., Ltd., Guangzhou, China

References

[1] Almeida AD, Santos B, Paolo B, Quicheron M. Solid state lighting review-potential and challenges in Europe. Renewable and Sustainable Energy Reviews. 2014;**34**:30-48

[2] Park DH, Lee DB, Seo ER, Park YJ. A parametric study on heat dissipation from a LED-lamp. Applied Thermal Engineering. 2016;**108**:1261-1267

[3] Tang Y, Ding X, Yu B, Li Z, Liu B. A high power LED device with chips directly mounted on heat pipes. Applied Thermal Engineering. 2014;**66**:632-639

[4] Deng Y, Liu J. A liquid metal cooling system for the thermal management of high power LEDs. International Communications in Heat and Mass Transfer. 2010;**37**:788-791

[5] Lin Z, Wang S, Huo J, Hu Y, Chen J, Zhang W, Lee E. Heat transfer characteristics and LED heat sink application of aluminum plate oscillating heat pipes. Applied Thermal Engineering. 2011;**31**:2221-2229

[6] Shin DH, Baek SH, Ko HS. Development of heat sink with ionic wind for LED cooling. International Journal of Heat and Mass Transfer. 2016;**93**:516-528

[7] Wang J, Cai Y-X, Bao W-W, Li H-X, Liu Q. Experimental study of high power LEDs heat dissipation based on corona discharge. Applied Thermal Engineering. 2016;**98**:420-429

[8] Yung KC, Liem H, Choy HS. Heat transfer analysis of a high-brightness LED array on PCB under different placement configurations. International Communications in Heat and Mass Transfer. 2014;**53**:79-86

[9] Park S-J, Jang D, Yook S-J, Lee K-S. Optimization of a chimney design for cooling efficiency of a radial heat sink in a LED downlight. Energy Conversion and Management. 2016;**114**:180-187

[10] Park S-J, Jang D, Lee K-S. Thermal performance and orientation effect of an inclined cross-cut cylindrical heat sink for LED light bulbs. International Journal of Heat and Mass Transfer. 2016;**103**:1371-1377

[11] Xiao C, Liao H, Wang Y, Li J, Zhu W. A novel automated heat-pipe cooling device for high-power LEDs. Applied Thermal Engineering. 2017;**111**:1320-1329

[12] Sömen KF, Yürüklü E, Yamankaradeniz N. Computational thermal analysis of cylindrical fin design parameters and a new methodology for defining fin structure in LED automobile headlamp cooling applications. Applied Thermal Engineering. 2016;**94**:534-542

[13] Jeng T-M, Tzeng S-C, Lin Y-C. Experimental study of heat transfer enhancement of inserted LED lamp by the closed-cell aluminum-foam ceiling. International Communications in Heat and Mass Transfer. 2015;**66**:233-239

[14] Wu Y, Tang Y, Li Z, Ding X, Yuan W, Zhao X, Binhai Y. Experimental investigation of a PCM-HP heat sink on its thermal performance and anti-thermal-shock capacity for high-power LEDs. Applied Thermal Engineering. 2016;**108**:192-203

[15] Sufian SF, Abdullah MZ. Heat transfer enhancement of LEDs with a combination of piezoelectric fans and a heat sink. Microelectronics Reliability. 2017;**68**:39-50

[16] Zhao X-J, Cai Y-X, Wang J, Li X-H, Zhang C. Thermal model design and analysis of the high-power LED automotive headlight cooling device. Applied Thermal Engineering. 2015;**75**:248-258

[17] Kang X, Wang Y, Huang Q, Cui Y, Wang C, Wen C, Fan J. Phase-change immersion cooling high power light emitting diodes and heat transfer improvement. Microelectronics Reliability. 2017;**79**:257-264

[18] Ahn B-L, Yoo S, Kim J, Jeong H, Leigh S-B, Jang C-Y. Thermal management of LED lighting integrated with HVAC systems in office buildings. Energy and Buildings. 2016;**127**:1159-1170

[19] Ahn B-L, Jang C-Y, Leigh S-B, Yoo S, Jeong H. Effect of LED lighting on the cooling and heating loads in office buildings. Applied Energy. 2014;**113**:1484-1489

Chapter 5

Development of Modular Cooling for Water-Cooled Photovoltaic Plant in Real Scale

Vinícius Oliveira da Silva,
Miguel Edgar Morales Udaeta,
André Luiz Veiga Gimenes,
Antônio Celso de Abreu Junior,
Angélica Luana Linhares and
Pascoal Henrique da Costa Rigolin

Additional information is available at the end of the chapter

http://dx.doi.org/10.5772/intechopen.79101

Abstract

This chapter evaluates module architectures and units of photovoltaic cooling systems, aiming to determine, select and design a modular system that can be applied in a real-scale photovoltaic power plant (PVPP) in order to enhance the yields of electricity production (entitled cooled photovoltaic plant). An analysis of the local climatic, geographic and solar conditions as well as construction, operational and maintenance aspects was carried out. Worldwide, there are three main types of cooled photovoltaic systems: PVT liquid and air collectors, PV ventilated with heat recovery and non-PVT systems. Based on the local weather conditions (tropical warm and dry) with both temperature and solar irradiation index being high, it results the PVT-liquid system to be more suitable in a scenario with available cooling fluid. We conclude that the best design and arrangement of the cooling system are of the type coil and multiple channel because they permit better rates of heat exchange between the cooling fluid and the PV module.

Keywords: SFV, solar energy, photovoltaic/thermal, cooling photovoltaic system, PVPP

1. Introduction

Brazil is the fourth largest country in the world with 8.5 $MMkm^2$ [1]; therefore, it presents diverse climatic characteristics along its territory. Nevertheless, the annual mean global irradiation remains uniform, presenting a relative high average in the country and varying in a range of 4200–6700 kWh/m^2—these values are, for instance, greater than most of the values observed in Europe (900–1850 kWh/m^2). The lowest global solar irradiation value observed in Brazil (4250 kWh/m^2) occurs in the northern coast of the state of Santa Catarina due to the steady rates of rainfall distributed along the year, whereas the highest value pertains to the northern region of the state of Bahia (6500 kWh/m^2), near the state of Piaui, because this region presents the lowest rate of cloud cover in Brazil [2].

As energy production (thermal or electrical) entails the increase of temperature of the equipment, directly affecting its efficiency, lifetime and pollutant emissions, cooling systems are employed in order to reduce the temperature of the system, such as radiators and water pumps employed in rotary machine of energy production (controlling overheating and NOx emissions of internal combustion engines Diesel and Otto), heat exchangers of a great variety used in hydraulic and gas turbines, which guarantee the maintenance of pillow blocks and shovels, among others [3, 4]. The photovoltaic modules (PMs), which are basically stationary solid equipment, are damaged with the effects of high temperatures during their operation because they absorb 80% of the solar radiation, which in turn only 5–20% is converted into electricity (depending on the employed technology) [5]. The rest of the solar radiation is then converted into heat, causing the modules to reach temperatures 35°C higher than the ambient temperature [6]; in some regions, these modules might reach temperatures higher than 90°C, causing malfunction or even permanent damage [7]. Alternatively, the photovoltaic thermal (PVT) system, which extracts the heat produced at the PV module through a thermal absorber system, might replace the traditional PV systems, improving its performance.

The aim of this work is to determine a model of cooling technology for photovoltaic power plants (PVPPs) by means of analysing both established and on-development cooling systems for PV modules. Therefore, we intend to improve the energy performance of photovoltaic power stations.

In these terms, the methodology considers the following steps:

1. study of local and geographic factors of the region where the photovoltaic power plant (PVPP) is located;

2. study of the factors that impact the increase of temperature on the photovoltaic (PV) modules and therefore impacting electricity production;

3. extensive study of the technical and scientific basis of the state of the art of cooling technologies for PV modules aiming at the understanding of the factors, which influence its functioning and performance. This step shall gather information regarding the materials employed on the construction of the selected technologies, methods and processes of construction, operation, maintenance and quantitative measurement;

4. survey and analysis of the local water availability because the PVPP is located within an operating hydroelectric power plant (HPP) and

5. evaluation of the cooling technologies in order to support the selection of the most appropriate one, taking into consideration the local characteristics.

2. Climatic aspects of the region for verifying energy application

The Northwest region of the State of Sao Paulo presents the highest score of mean solar radiation of the state, with a mean annual value of 5.520 kWh/m^2 and daily maximum value of 5.970 and 6.672 kWh/m^2 day, respectively, in the summer and spring, which lie close to the maximum values observed in the country [8]. The installation site of the photovoltaic power plant (PVPP), where the cooling systems for the PV modules are installed [9], presents an average annual solar radiation of 5.5 kWh/m^2 day in the summer and minimum average value of 4.6 kWh/m^2 day in the autumn [8], as shown in **Figure 1**.

According to Köppen classification, the city of Ilha Solteira presents warm and dry climate [10], with an average annual atmospheric temperature of 24.8°C and the maximum and minimum average annual temperatures of 33.0 and 13.1°C, respectively. Ilha Solteira has an annual rainfall index of 1309.4 mm, varying from 288.3 mm during the dry season (from April to September with an average monthly index of 48.1 mm) to 1.021 mm during the wet season (from October to March with an average monthly index of 170.2 mm, comprising 77% of the total rainfall in a year). The minimum rainfall index is observed in August (22.8 mm), whereas the maximum one is observed in January (225.9 mm). Regarding atmospheric temperatures, the dry season presents average value and maximum average value of 22.8 and 30.0°C, respectively, whereas the wet season registers 26.9 and 32.5°C. The highest registered temperature was 42.0°C, which was registered on four different days, and the highest average daily temperature was of 34.9°C, with minimum of 29.3°C and maximum of 38.9°C on that day [11].

Figure 2 shows rainfall indices (mm) and daylight (h/day) for the period 2004–2013. The graphic reveals that the minimum values registered for sunshine duration coincide with

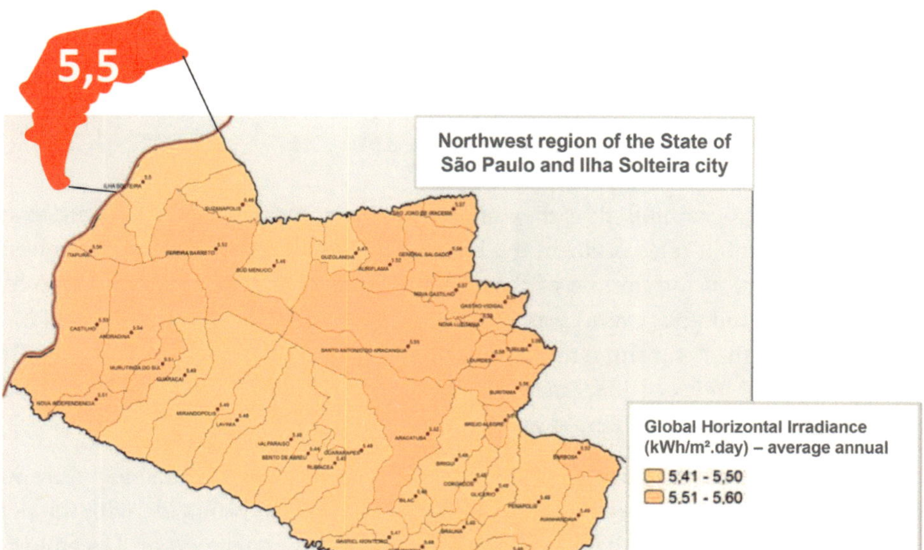

Figure 1. Global horizontal irradiance (kWh/m^2) in the region on Ilha Solteira city [8].

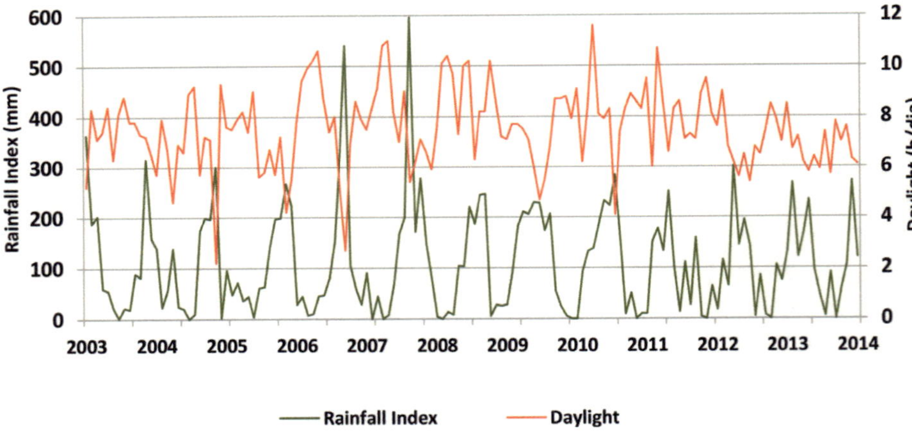

Figure 2. Rainfall index and monthly daylight for the period 2004–2013.

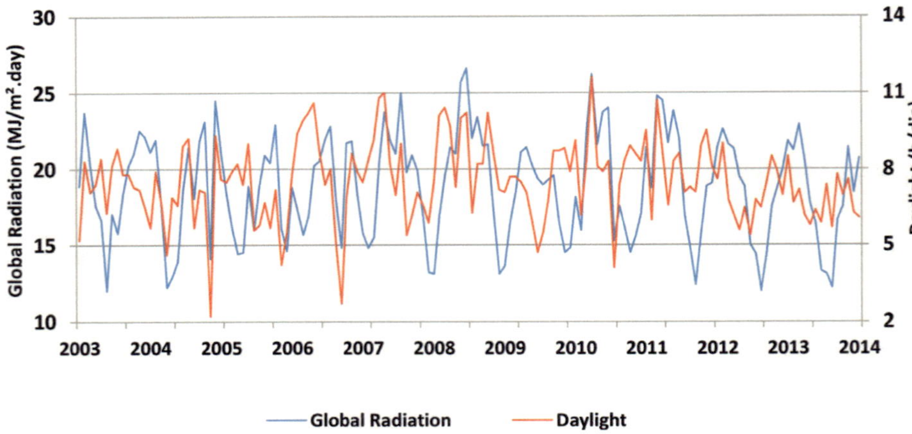

Figure 3. Mean global radiation and monthly daylight for the period 2004–2013.

the maximum values of rainfall, thus they are inversely proportional and oscillate along the year. The highest rainfall rates occur in the beginning of the year, whereas the highest sunshine duration occurs in the middle of the year. It is also possible to note that the highest rainfall index registered (596.1 mm) was followed by a monthly sunshine duration of 5.4 h/day, whereas the highest sunshine duration (11.6 h/day) registered in November 2010 was followed by a rainfall index of 138.9 mm. Overall, the mean monthly sunshine duration and rainfall indices for the analysed period are 7.7 h/day and 127.6 mm, respectively.

The global radiation (MJ/m^2 day) is directly proportional to the sunshine duration, as described in **Figure 3**, and the peaks of maximum global radiation coincide with the peaks of sunshine. The highest mean global radiation (26.6 MJ/m^2 day registered on December 2008) corresponds to the maximum mean atmospheric temperature of the same year (26.7°C), presenting a mean monthly sunshine duration of 10.2 h/day. At the period of highest sunshine (11.6 h/day on November 2010), the mean global radiation was 26.2 MJ/m^2 dia, corresponding to the second highest value for the period. The mean average global radiation for the whole period is 19.1 MJ/m^2 day. Furthermore, the graphic shows that the oscillations registered for the maximum global radiation are smoother than the ones observed for the minimum ones, thus the minimum values vary more than the maximum ones. Another important characteristic to be noted is that the mean global radiation values along the period lie closer to the maximum values than to the minimum ones, demonstrating that conditions of high radiation are more present in the region. The average minimum and maximum global radiation for the period is, respectively, 9.4 and 24.9 MJ/m^2 day, reaching a peak of 35.4 MJ/m^2 day observed on 4 December 2008, with a corresponding sunshine duration of 15.8 h/day.

3. Analysis of the effect of temperature on PV modules

The temperature is a very important parameter on PV modules due to its influence on the behaviour of a PV system. The PV cells heat up due to their exposure to the solar rays, modifying the efficiency of the system and the output energy, because high temperatures decrease the produced voltage, and inversely, higher voltages are produced in lower temperatures [12]. Any kind of PV system shall include a correction factor due to the effect of temperature [13]. This effect is a product of a natural characteristic of the silicon [12] because part of the absorbed solar rays is not converted into electric energy but dissipated in the form of heat. This is the reason for the PV cells and modules to present higher temperatures than the ambient while operating [14]. The highest temperatures on a PV system are observed at its inferior surface due to the higher thermal conductivity of the silicon that forms the PV cell, compared to the polymer material used at the superior surface [15]. This phenomenon is shown in **Figure 4**, presenting the difference of temperature among the superior and the inferior surfaces of 8.1°C [3, 4].

The voltage drop caused by the increase in the temperature is a result of the shrinkage of the semiconductor due to the band gap, which directly influences the open-circuit voltage (Vca), which in turn decreases following the voltage drop of the p-n junction. This effect can be explained by the thermal voltage, q/kT. Thus, the PV cells contain a negative temperature coefficient for the Vca. Moreover, lower output power given a constant photocurrent results from the charge carries, which are released with lower potential [16]. As the temperature increases, followed by an opening of the band and thus shrinking the semiconductors, more incident energy is absorbed, leading to a raise in the charge carriers from the valence band to the conduction band [12]. As a result, a larger photocurrent is observed, producing a larger short-circuit current for a given insulation, and PV cells have a positive temperature coefficient Icc [16].

According to Makrides et al. [17], the place in the PV cells, where the highest increase in temperature is observed, is at the main bus bar; a conductor line that enables the interconnection, capture and conduction of electrons from the secondary bars, which, as the primary ones, is a parallel conductor line responsible for the capture of electrons produced at the valence band of the PV cells. According to Prieb [18], as the silicon module temperature increases, the voltage decreases at a rate of −2.2 mV/°C and the short-circuit current increases at a rate of

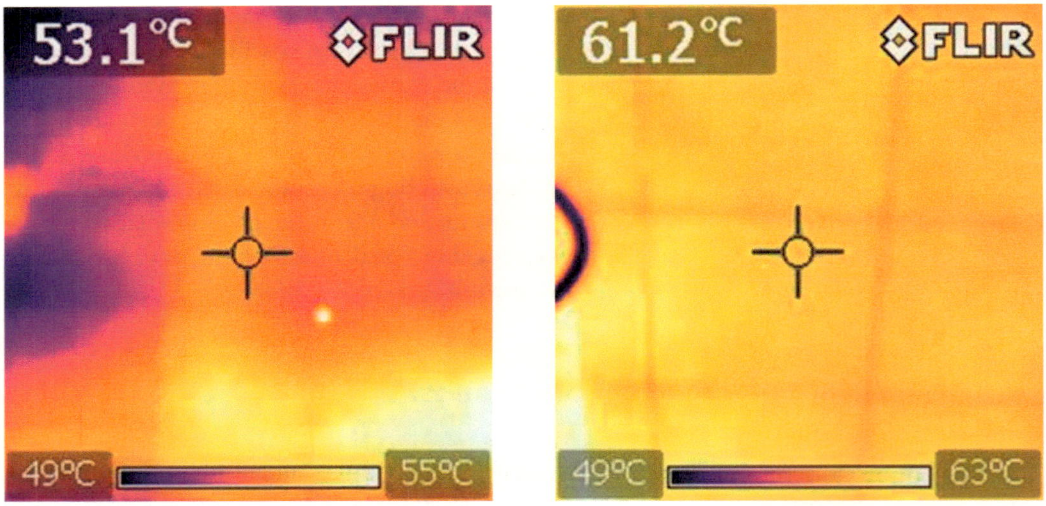

Superior surface Inferior surface

Figure 4. Infrared image of the superior (left one) and inferior (right one) surfaces of a PV module [3].

0.06%/°C, such increase rebounds on the power, but it is insufficient to compensate the loss of power due to the voltage drop. In summary, variation of temperature results in a corresponding variation of operational levels of maximum power extraction.

The energy conversion efficiency (η) of a PV cell is the percentage between power converted and power collected, when a PV cell is connected to an electrical circuit. This term can be calculated using the point of maximum power (P_{max}) divided by the input light irradiance (E in W/m^2) over a PV cell surface under standard test conditions (A_c). In this way, it is possible to understand the decrease of η influenced by temperature increase, because such a decrease in the output power is directly proportional to the output voltage, which in turn is reversely proportional to temperature. Thus, just as the voltage, the efficiency of a PV module decreases with the increase of temperature.

4. Framework of cooling systems for photovoltaic modules

In photovoltaic thermal (PVT) systems, solar energy is converted into both heat, just as in conventional solar thermal collectors, and electricity. In other words, PVT systems consolidate the electricity production of PV systems and the heat production of solar thermal collectors (see **Figure 5**). The real conversion occurs at the absorber, but part of this energy dissipates to the environment through radiation and convection. Systems where the absorber is in direct contact with other means are called uncovered PVT. In this case, the heat loss to the environment is considerable, and both temperature and thermal efficiency are relatively low. The uncovered PVT systems are recommended in cases when lower heating temperatures for

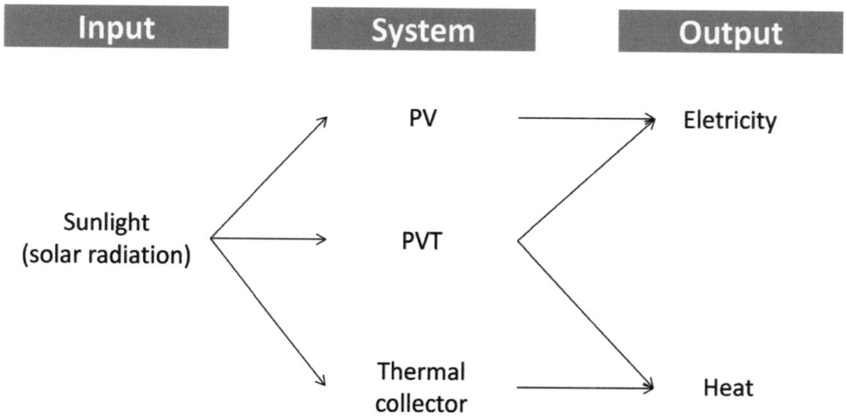

Figure 5. Comparison between the types of technology and the corresponding energy produced [19].

water are demanded or in order to improve the electric conversion yields of PV modules. The other kind of system is called covered PVT, and it is characterized by a cover, usually transparent, placed around the absorber. This cover transmits around 90% of solar incident radiation, depending on the material used. On most of the cases, the 10% reduction is less important than the effect of thermal insulation the cover achieves, improving the overall thermal efficiency of the system [6].

The parameters of a PVT project vary substantially according to the type of application, which can aim at the use of hot water for pre-heating ventilation air or at cooling the PV systems. Heating and electricity demands can be satisfied by the appropriate PVT system, and its technical viability is documented according to Hasan and Sumathy [19].

There are many types and forms of PVT systems, which vary according to the photovoltaic module, the type of working fluid, the amount of radiation and geographic location of the facility, and other system specifications (such as flat plate or concentrator). The actual classification of PVT systems varies in the literature—the working fluid (commonly air and water) is the most common attribute used for classification purposes.

Compared to the air-type PVT, the water-type PVT systems are more efficient in terms of heat removal from PV [20], due to the higher thermal conductivity of the water, resulting in a higher heat transfer from the PV modules to the working fluid. Thus, a balance between thermal and electric yields has to be taken into account while selecting the appropriate PVT system.

4.1. Water-based PVT collector

Water-based PVT collectors consist of a heat absorber, either a cooling coil or a series of parallel tubes coupled at the inferior surface of the PV laminate or glued with an epoxy adhesive material through which a fluid flows, cooling down the module. Moreover, the absorber might contain a thermal insulator in order to reduce thermal losses between the water and the ambient—in contrast, in case heat removal from the PV module is intended, the thermal insulator is not added to the absorber. This type of cooling is accomplished by forced convection in which the fluid removes the heat from the module, increasing the system's electrical efficiency of 4–5% [3].

There are several types of configuration for these collectors, but in all cases, a pump has to be installed for water circulation purpose. The four major models found in the literature are: sheet-and-tube PVT collectors, channel PVT collectors, free flow PVT collectors and two-absorber PVT collectors. According to Zondag et al. [21], in order to improve the efficiency of the collector, it is preferable to design the water flow to occur below the PV module. However, in terms of the simplicity of the structure and the overall performance, a sheet-and-tube PVT collector is recommended. Additionally, in order to maximize heat transfer, the mean distance between the heat generation and the collectors should be minimum. Another suggestion regarding the area in which the liquid flows on the PV cell, this shall be as large as possible in order to cover the whole surface [22]. For this application, van Helden et al. [6] suggested a configuration with multiple channels that allow PV cells arranged in series to be cooled equally.

The simplest way to produce a PVT collector is to take a standard PV and integrate it into a thermal collector. However, this configuration exposes the PV module to the ambient, particularly to humidity, and therefore hampering practical applications. Furthermore, problems related to the electrical insulation may occur, such as the increase of thermal resistance between the PV laminate and the absorber due to the emergence of air layers—result from the low thermal conductivity triggered by irregularities at the inferior p-Si modules [21].

In order to increase electrical resistance between the PV cells and the absorber, a more advanced technique shall be applied, avoiding temperature increase and electrical loss. This technique is based on PVT modules that contain laminate photovoltaic cells followed by electrical insulation and finally an absorber. However, problems such as the deformation of the PVT laminate due to the thermal expansion between the glass surface and the metal laminate of the collector occur. In order to avoid this problem, a tedlar laminate can substitute the glass at the inferior surface. It is important to note that this substitution requires a sufficient rigid absorber that enables the cells to be supported [21]. Usually, PVT modules produced in strips with a superior plastic layer have been used. In this case, the support is guaranteed by a cooper tube absorber installed at the centre of the strips. A variation of this design uses the borders of a galvanized sheet as the support of the PVT modules. The border of the galvanized sheets is bent, providing the basin for the cooper tubes, which are then welded together. Sheet-and-tube absorbers present technical restrictions at the construction phase because the tubes are located underneath the absorber, leading to difficulties at the lamination process. Overcoming this problem is extremely complex because the high temperatures of the welder hamper the encapsulation process and diminish the reliability of the heat transfer. The utilization of an aluminium sheet, which is introduced by a cladding process, may overcome such limitation [23].

It is worthwhile to mention that, independent of the selected technique, the encapsulant material has to be resistant to temperatures as high as 130°C, and particular attention to the optical properties of the PV cells has to be paid [24]. An alternative way to avoid the complications caused by high temperatures consists of applying silicon. However, this option presents the disadvantage of diminishing the efficiency of thermal exchanges in case an air layer is accidentally produced.

4.2. PVT-air collector and ventilated PV with heat recovery

PVT-air collectors are similar to a standard hot air collector with a PV laminate applied at the superior cover of the air channel. PVT-air collectors are cheaper than the liquid ones due to its flexibility and easy conversion properties. They can be constructed with or without glass, and, in general, they are recommended in the cases when the user has a demand for hot air specially to be directly applied for heating purposes. This application is restricted to places where there is a demand for hot air usually applied for pre-heating in building systems, where the temperature demand ranges from 15 to 25°C [19].

As the heat transfer in a PVT-air system is more critical than in the liquid ones, it is important to adopt an adequate heat transfer model. Due to heat transfer effects at the entrance, the Nusselt number (Nu) in PVT-air systems varies 10% along the entrance length for a given sufficient large channel, and therefore, the hydraulic diameter shall be twice the size of the height of the channel. The impact of the air flow induced by buoyancy and heat transfer through a vertical channel induces an increased velocity of the heat flux non-uniformly inside the duct. This effect is directly connected to the form of the exit side of the duct. Briefly, in order to enhance the heat transfer in a PVT-air system, it is necessary to control its induced flow and buoyancy. Another option extensively applied in experimental research consists of increasing both the turbulence in the flow channel and the heat transfer surface area [19].

In terms of constructive aspects of the fixation of the cooling system of the PV modules and the corresponding used materials, PVT-air systems are similar to PVT-liquid ones. The usage of air as the heat transport medium as opposed to other working fluids, for example, water, presents some advantages such as the non-freezing property of the collector fluid (important factor when applied in low temperature regions), no risk in case of leakage, and the input is freely and permanently available. This solution is indicated for regions where the access to water is restrict or economically unviable. Nevertheless, some disadvantages shall be taken into consideration, such as low heat capacity and low heat conductivity, which result in low heat transfer. Thus, this configuration is not recommended when the cooling of the PV modules is aimed (there is considerable heat loss), and, in case of a passive systems, without an air injection system, the PVT-air system may result in low yields due to the low density and thermal conductivity properties of the air.

The difference between a ventilated PV with heat recovery and PVT collectors reside in the fact that, generally, a PVT system is projected for a specific application (e.g., building), lacking a standardized production system. The similarities between both the systems cause practitioners to often select the wrong option for the project. This scenario may change in the short term once various institutes and manufacturers, specially in Europe, are channelling efforts towards standardization.

Conventional PV modules for facades and roofs, where frequently air incidence is present at the inferior surface of the modules due to its inclination, enable the air to cool the PV by means of natural convection. This heat can be recovered, and, in this case, the PV is considered to act as a PVT collector. Due to its easier construction and operation, PVT with heat recovery systems is extensively studied as an alternative solution to "Building Integrated Photovoltaic" (BIPV). This system can operate either during the winter, providing heat, or during the summer as active cooling systems [25]. An extensive research carried out by Bazilian et al. [26] proved that this type of system is more suitable for low temperature applications. Moreover, he observed that building integrated systems could represent a cohesive project and become a good solution for providing energy in buildings, pointing the necessity to further increase the research in the area in order to make these systems commercially available. Another research [27] also showed that BIVP collectors are more suitable for low temperature ambient. This system, if applied at facades, can not only provide the electricity gains but also protect the building against solar radiation, reducing the thermal cooling load and providing the heat for internal uses in the building. In the absence of direct demand of heat, this can be utilized to induce a pressure difference, supporting the ventilation system. Additionally, the PV modules may substitute facade cladding materials [19].

4.3. Cooling system of PV module (non-PVT)

The cooling system of PV modules non-PVT consists of water injection on the module's superior surface. As the water flows on the surface, it removes the heat stored at the module through conductive and convective heat transfer. This system is simple and is composed of a PV module, a pump and a water storage tank. The water injection is done over the module's superior surface through various orifices installed along a tube, guarantying homogeneous heat exchanges along the whole surface.

Results obtained by [28] showed that the convective heat exchange between the cooling water and the module's superior surface lead to increases of 15% in electricity production during peak solar radiation. Moreover, the results indicated that a 5% increase in the energy output can be obtained during dry and warm seasons. In another experiment, Odeh et al. [29] designed a system in which tubes were incorporated in order to allow the water to flow on the module's surface based on gravity, improving the yields related to the water pumping system. The test was carried out in different cities of Australia (Sydney, Perth and Darwin) in order to consider different climatic conditions. The results showed that the increase in output power of the systems varied in a range of 4–10% when the cooling system was operating. Around 50% of the increase is attributed to the direct cooling effect resulted from the contact between the water and the surface, whereas the other 50% comes from the increase of solar radiation due to the refraction of the light beam on the water layer.

5. Important aspects of the water supply for PV systems

The PVPP to be tested in this work is located within the facilities of a hydroelectric power plant (HPP), enabling the ease of access to the cooling water. The use of the water is possible in two different ways: either it is done at the end of the hydroelectrical energy production process, after the water runs through the turbines, or it is directly applied at the reservoir, before the energy production phase, with the condition that the water has to be reintroduced in the system, either in the reservoir or at any step of the hydroelectrical energy production. After analysing numerous configurations, the following two available options are analysed and discussed.

5.1. Output water from the cooling of the pillow blocks

Cooling water from the pillow blocks is permanent available along the year only at the operation period in which the cooling machines operate. Before the water enters the heat exchanger of the pillow blocks, it passes through a chemical process (chlorination) aimed to control mussel fouling. In order to seize the water at this step of the process, several interventions at the structure of the dam would have to be carried out, such as drilling on the walls and ceilings of the machine rooms, redeployment of electrical conduits and lubricants located at the dam's internal channels, power down and drilling of the main discharge line from the heat exchanger of four turbines and construction of a 700-m pipeline up to the dam's rockfill plus an enlargement of 77 m from the rockfill to the PVPP. During the summer, the water temperature measured at this site presents a maximum value of 39°C and an annual average of 32°C and, at the coldest period, a maximum of 29°C and an average of 26°C [3].

- Advantage: chlorinated water, avoiding the necessity of treatment water plant for the cooling of the PVPP; output water, avoiding the necessity of reintroducing the water into any other step of the HPP; no pumping process required.
- Disadvantage: several physical interventions at the dam's structure; power down of two generators, if these are operating during the implementation of the cooling system; redeployment of electrical conduits and lubricants used at the operation of the generators; construction and enlargement of pipes; high water temperature; water supply dependent on the operation of the generators.

5.2. Rockfill water from the HPP

The rockfill raw water from the HPP is collected at the bottom of the dam; therefore, its temperature is more constant along the year, with an average value of 24°C, a minimum of 19°C in the winter and a maximum of 29°C in the summer. The rockfill line runs over less than 500 m between the exit of the HPP's internal facilities and the administrative parking zone. The distance between the ending point of the rockfill line and the PVPP facility is of 77 meters. Among others, this water has been used for cleaning, gardening, hydrant backup and external supply. These conditions enable the permanent and continuous supply of water along the whole year [3].

1. Advantages: no necessity for interventions at the dam's structure; no necessity for shutting down any HPP's operation; ending process water, avoiding the necessity to return it; no pumping system required; relatively small extension of the pipeline; permanent water flow along the year.
2. Disadvantage: implementation of a chlorination system.

6. Conclusion

The PVPP's facility site presents a dry and warm weather, with an average temperature of 24.8°C and a rainfall index of 1.309 mm. The rain season occurs between October and March with 77% of the annual rainfall lying within this period, whereas the dry season occurs between April and September. The mean global solar radiation and the mean sunshine duration are, respectively, 19.1 MJ/m^2 day and 7.7 h/day. Furthermore, the highest historic temperature (42°C) was recorded within the rainy season. These characteristics demonstrate high sunshine duration and global radiation along the year.

Based on the analysis of this report, the liquid PVT without thermal insulation is the most recommended solution. The fact that the PV modules present inferior temperature 8.1°C higher than the superior one (filed measurements), the absence of waste water (closed system) and the high demand for cooling the PV modules for energy production corroborates the adopted solution. Another advantage of liquid PVT systems is that it enables the use of standard PV modules, which present high thermal exchange rates, being highly recommended for warm regions. It is important to mention that three configurations are possible among the liquid PVT system, two of those regard sheet-and-tube (coil) and the other is of the type multiple channel.

Moreover, based on qualitative analysis of the local conditions observed on the site, the most suitable source of cooling fluid is the water from the rockfill line of the HPP, because of the proximity between the existing line and the PVPP area, permanent and continuous water supply along the year, no pumping required and, most importantly, the absence of structural interventions at the dam.

Acknowledgements

To CESP for funding the P&D ANEEL PE-0061-0037/2012, enabling the accomplishment of this chapter and especially for the engineer Diego Biaseto Bernhard of the company Flue8 that due to its engagement and experience solved all technical and constructive aspects of the cooling units.

Author details

Vinícius Oliveira da Silva[1*], Miguel Edgar Morales Udaeta[1], André Luiz Veiga Gimenes[1], Antônio Celso de Abreu Junior[1,2], Angélica Luana Linhares[1] and Pascoal Henrique da Costa Rigolin[1,3]

*Address all correspondence to: vinicius.oliveira.silva@usp.br

1 Energy Group at the Engineering Department of Electrical Energy and Automation of the Polytechnic School, University of São Paulo (GEPEA/EPUSP), São Paulo, Brazil

2 Department of Energy and Mining of the State of São Paulo, São Paulo, Brazil

3 Center for Urban Energy, Ryerson University, Toronto, Canada

References

[1] IBGE (Brasil). Instituto Brasileiro de Geografia e Estatística. DOU n° 16 de 23/01/2013. Brasília: IBGE; 2013

[2] INPE (Brasil), Pereira EB, Martins FR, Abreu SL, Rüther R. Atlas brasileiro de energia solar (ABES). São José dos Campos: INPE; 2006

[3] Silva VO. Estudo e modelagem da arquitetura modular de uma Usina Fotovoltaica arrefecida com protótipo de verificação (Mestrado). São Paulo: Escola Politécnica da Universidade de São Paulo; 2015. 167 p

[4] Gimenes ALV, Linhares AL, Abreu Junior AC, Bernhard DB, et al. Desenvolvimento experimental de um sistema de resfriamento para geradores fotovoltaicos em escala real. Espacios. 2018;**39**(04):33

[5] Silva VO, Gimenes ALV, Galvão LCR, Udaeta MEM. Study, verification and selection of cooling system model for FV modules with verification prototype. In: European Photovoltaic Solar Energy Conference and Exhibition (EU FVSEC). Hamburg; 2015. pp. 2153-2158

[6] van Helden WGJ, van Zolingen RJC, Zondag HA. PV thermal systems: PV panels supplying renewable electricity and heat. Progress in Photovoltaics: Research and Applications. 2004;**12**:415-426

[7] Oh J, Samy G, Mani T. Temperature testing and analysis of PV modules per ANSI/UL 1703 and IEC 61730 Standards. In: Conference Record of the IEEE Photovoltaic Specialists Conference, Program—35th IEEE Photovoltaic Specialists Conference, PVSC 2010. 2010. pp. 984-988

[8] SEE (Brasil). Secretaria de Energia do Estado de São Paulo. Energia Solar Paulista: Levantamento do Potencial, Secretaria de Energia, Subsecretaria de Energias Renováveis, São Paulo; 2006

[9] Aneel (Brasil). Desenvolvimento de Equipamento de Usina Solar Fotovoltaica com Sistema Ativo de Arrefecimento para maior Rendimento na Geração de Eletricidade. Projeto Estratégico de Pesquisa e Desenvolvimento ANEEL PE-0061-0037/2012. Programa de P&D da Companhia Energética de São Paulo, CESP; 2013

[10] CEPAGRI (Brasil). Centro de Pesquisas Meteorológicas e Climáticas Aplicadas a Agricultura. Campinas, Brasil: UNICAMP; 2015. http://www.cpa.unicamp.br/outras-informacoes/clima_muni_236.html [Acessado: maio-2015]

[11] Unesp (Brasil). Universidade Estadual Paulista. 2013. http://clima.feis.unesp.br/recebe_formulario.php [Acessado: agosto-2013]

[12] Meral ME, Dinçer F. A review of the factors affecting operation and efficiency of photovoltaic based electricity generation systems. Van, Turkey: Yuzuncu Yil University, Department of Electrical and Electronics Engineering; 2011

[13] Krauter S. Increased electrical yield via water flow over the front of photovoltaic panels. Solar Energy Materials and Solar Cells. 2004;**82**:13-37

[14] TEE. Tools for Electrical engineering, e-learning. http://e-lee.ist.utl.pt

[15] Teo HG, Lee PS, Hawlader MNA. An active cooling system for photovoltaic modules. Applied Energy. 2012;**90**:309-315

[16] Shenck NS. Alternative Energy Systems. U.S. Naval Academy Lecture Readings; 2010

[17] Makrides G, Zinsser B, Georghiou GE, Schubert M, Werner JH. Temperature behavior of different photovoltaic systems installed in Cyprus and Germany. Solar Energy Materials & Solar Cells. 2009;**93**:1095-1099

[18] Prieb CWM. Desenvolvimento de um Sistema de Ensaio de Módulos Fotovoltaicos [Dissertação para Obtenção do Título de Mestre em Engenharia]. UFERGS; 2002

[19] Hasan MA, Sumathy K. Photovoltaic thermal module concepts and their performance analysis: A review. Renewable and Sustainable Energy Reviews. 2010;**14**:1845-1859

[20] Prakash J. Transient analysis of a photovoltaic thermal solar collector for cogeneration of electricity and hot air water. Energy Conversion and Management. 1994;**35**:967-972

[21] Zondag HA, De Vries DW, Van Helden WGJ, Van Zolingen RJC, Van Steenhoven AA. The yield of different combined PV-thermal collector designs. Solar Energy. 2003;**74**(3):253-269

[22] Silva VP, Gimenes ALV, Udaeta MEM. Development of a real-scale cooling module for a PV power plant. IET Renewable Power Generation. 2018;**12**(4):450-455

[23] Chow TT, He W, Ji J. Hybrid photovoltaic-thermosyphon water heating system for residential application. Solar Energy. 2006;**80**(3):298-306

[24] Elnozahy A, Rahman A, Ali A, et al. Performance of a PV module integrated with standalone building in hot arid areas as enhanced by surface cooling and cleaning. Energy and Buildings. 2015;**88**:100-109

[25] Eicker U, Fux V, Infield D, Li M. Heating and cooling of combined PV-solar air collectors facades. In: Proceedings of International Conference of 16th European PV Solar Energy. 2000. pp. 1836-1845

[26] Bazilian M, Leeders F, van der Ree BGC, Prasad D. Photovoltaic cogeneration in the built environment. Solar Energy. 2001;**71**:57-69

[27] Charalambous PG, Maidment GG, Kalogirou SA, Yiakoumetti K. Photovoltaic thermal (PV/T) collectors: A review. Applied Thermal Engineering. 2007;**27**:275-283

[28] Endecon Engineering: Energy Technology Development Division, Consultant Report, a Guide to Photovoltaic (PV) System Design and Installation. California Energy Commission; 2001. pp. 1-40

[29] Odeh S, Behnia M. Improving photovoltaic module efficiency using water cooling. Heat Transfer Engineering. 2009;**30**(6):499-505. DOI: 10.1080/01457630802529214

Chapter 6

Nucleate Pool Boiling Heat Transfer of Refrigerants Using Coated Surfaces

Ashok K. Dewangan, Anil Kumar and Ravi Kumar

Additional information is available at the end of the chapter

http://dx.doi.org/10.5772/intechopen.81864

Abstract

This work presents the experimental study of nucleated pool boiling heat transfer of R-134a and R-410A on a horizontal coated heating surface. The heating surface dimensions are 25.4 mm outer diameter and 116 mm effective length. The coated surfaces were fabricated by flame spraying technique. The copper powder was used as a coating material applied to the outer surface of copper tube. The experiments were performed for heat flux range of 5–50 kWm^{-2} at saturation temperature of 10°C. The heat transfer coefficients of both refrigerants demonstrated the same trends with applied heat flux increase and their magnitudes increases with increasing the value of applied heat flux. The present study also includes the effects of heat flux and coating parameter on boiling characteristics. The boiling heat transfer coefficient is enhanced by 1.9 times that of plain surface. An empirical correlation was also developed to predict the heat transfer coefficient with a mean error of 13%.

Keywords: pool boiling, heat transfer enhancement, coated surface, refrigerants, flame spraying

1. Introduction

Pool boiling characteristic has immense heat transfer applications because of the ability to remove enormous quantities of heat from heating surface with maintaining the lower temperature difference. This gives a reduced size of heat exchanger by enhancing the performance of equipments used in many industries such as refrigeration and air-conditioning industries, thermal power plants, process industries, and many other allied industries. In refrigeration and air-conditioning field, flooded evaporator is widely used as heat exchanger surface. The energy Industry demands for more enhancing surface and economic incentives as well, have spurred the development of methods to increase the heat transfer coefficients. The various types of surface modification are used to improve the performance of equipments. Among the various available surface coating methods, metallic coating is one of the appropriate coating material for use in pool boiling of refrigerants. This is considered for the basis on the overall consideration of enhanced performance, durability, the ease of manufacturing and application. Hence, a flame spraying technique was used to prepare the coated surfaces. The copper powder was used as a coating material applied over the surface of copper tube.

The impacts of Refrigeration and air conditioning system on environment are majorly supposed to release ozone-depleting refrigerants. The increase in emission of halogenated refrigerant used in refrigeration systems increases the concentration of greenhouse gases in the environment. As a result, adverse climatic changes being observed recently and contribute significantly to global warming. A reduction in GHG emissions can only be achieved by using alternative refrigerants. Thus to protect the earth surface from direct infrared radiation and to find solutions to socioeconomic favor for the mankind, further study in this area is indispensable.

In order to fill the gap caused by the phase out of CFCs, researches have been carried out extensively to find alternative refrigerants whose ozone depletion potential (ODP) is zero. Nowadays, design engineers have put remarkable efforts in designing efficient and compact systems thus reducing ODP. Therefore, R-134a can be better alternative refrigerant for CFCs because of its good thermodynamic properties as well as eco-friendly features. The refrigerant mixture R-410A is also a long-term alternative refrigerant with zero ODP for time being in developing countries. Due to the low temperature efficiency and lower discharge temperatures, favorable physical and transport properties, R410A are widely preferred in refrigeration and air conditioning applications. In present study, the refrigerants R-134a and R-410A were used as alternative refrigerants. A very few research works on pool boiling of refrigerants on coated surfaces are reported in the literatures.

Nowadays, many researches have focused on enhanced boiling heat transfer surface which fulfill the requirements of advance developments in energy generating equipments. Thome [1]; Webb [2]; and Bergles [3] were discussed the different enhancement techniques for fabricating the heating surfaces. Active, passive and compound were considered as enhancement techniques. A passive technique does not require any external power source and also its fabrication process of heating surface is easy and economical. The metallic coated surface is prepared by the passive techniques for enhancing the boiling heat transfer. The boiling of new refrigerants on metallic coated surfaces has been studied by very few researchers [4–18]. However, the experimental work on metallic coated surfaces and their impact on boiling heat transfer are still indispensable. The boiling heat transfer of refrigerants on these surfaces generates more active nucleation sites and it depends on the properties of refrigerants and surface geometry. The main objective of the present study is to conduct the pool boiling experiments of refrigerants R-134a and R-410A in metallic coated surfaces. This study provides the data to the refrigeration industry for the design of efficient heating surface. In addition, parametric study on the boiling characteristics is analyzed which determines the probability of flooding the reentrant cavities and the amount of superheat required for bubble growth.

2. Experimental facility and procedure

Figure 1 represents the experimental setup for the boiling of refrigerants on coated surfaces. This setup consists of boiling vessel, power supply arrangement, condensing loop and test section. The sealed cylindrical boiling vessel was fabricated with a 490 mm long stainless steel pipe of 150 mm internal diameter. It is closed at both ends with flange of same material. The bottom cover of boiling vessel contains fitting to mount a pipe with a valve to charge or drain out refrigerant from the vessel as and when required. This bottom cover has also fitting for preheater to control the saturation pressure of the system. Two inspection windows were welded at diametrically opposite side position of the boiling vessel body for visual observation of bubble dynamics on and near the tube surface. The boiling vessel was well insulated with glass wool to ensure adiabatic condition. A powder flame spraying method is applied to fabricate heating surface where copper powder used as coating material. The details of the test section along with the cartridge heater shown in **Figure 2**. The test sections were heated by cartridge heaters. Each cartridge heater having 16.5 mm diameter and actual heated length of 116 mm, and was inserted into the copper tube. K-type chromel-alumel thermocouples were embedded in the circumferential position of the tube to measure the wall temperatures. Four holes at top, two sides and bottom positions were made circumferentially in the wall thickness of test tube. Four thermocouple probes were embedded to

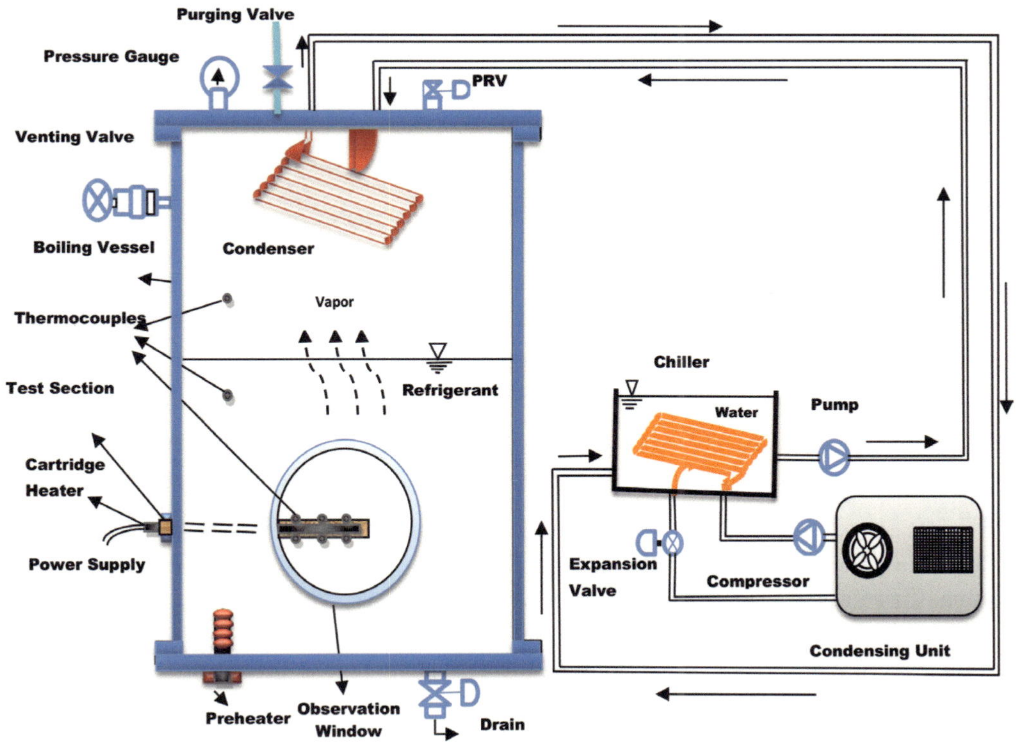

Figure 1. Schematic diagram of experimental set up.

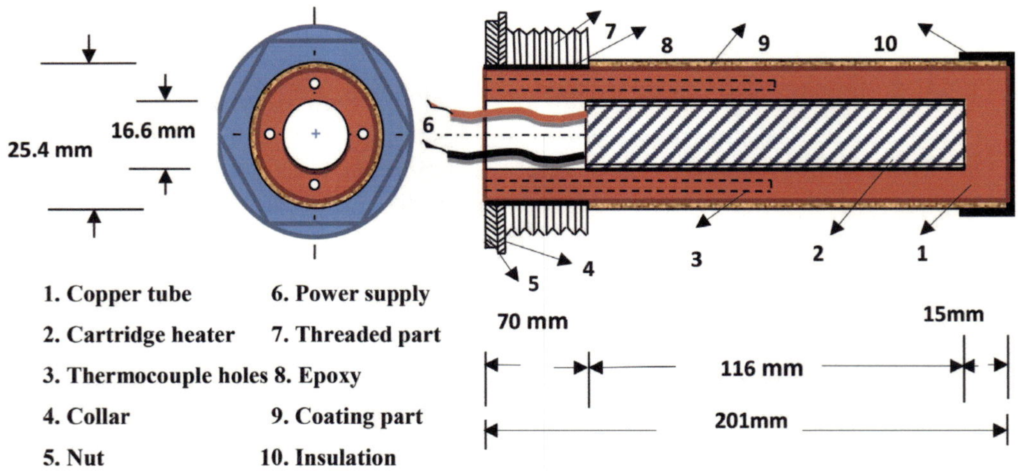

Figure 2. Test section along with the heating arrangement.

measure the temperature at each hole and the average of these temperatures indicates the wall temperatures. Two thermocouple probes were also inserted through the top cover of the boiling vessel at suitable positions to measure the liquid pool temperature on opposite sides of heating tube.

The refrigerant vapor produced in the boiling vessel get condensed in an internal water cooled condenser and returned to the vessel. The internal condenser was mounted vertically below the top cover of the vessel to ensure return of the condensate by gravity to liquid pool. An external chiller with an accurate temperature controller was used to condense to maintain the pool temperature to 10°C. The heat transfer experiments were conducted for boiling of three refrigerants R-134a, and R-410A on plain and coated tubes at saturation temperature of 10°C. The following procedure was adopted for pool boiling of given refrigerants on coated surfaces:

- Before each test, the boiling vessel and fabricated test surfaces were cleaned with acetone, chlorinol and water.
- A pressure and vacuum gauges were used to check the leakage inside the boiling setup maintaining a pressure of 2.0 MPa and a vacuum of 60 cm of Hg for 24 hours. When the system is ensured from leakage then refrigerant was filled into boiling vessel as vapor form to the level of 35 mm above the boiling surface.
- To remove the non-condensable gases and air from the boiling vessel through purging valve, the refrigerant was heated at 50 kWm^{-2} for 1 hour. When the complete removal of air from the boiling vessel was ensured, the chiller was started 3 hours prior to beginning of the experiments.
- The cartridge heater was used to supply the power to test surface in the range of 5–50 kWm^{-2} for both increasing and decreasing heat flux levels. This was done for the purpose of avoiding a boiling hysteresis. The saturation temperature of the refrigerant was maintained at 10°C within 0.2°C temperature fluctuation.

Test tubes	Coating thickness, t_c (μm)	Porosity, ε (%)	Mean pore diameter, d_{mp} (μm)
C-1	42	11.03	2.58
C-2	95	13.1	2.52
C-3	151	8.5	1.75
C-4	271	13.8	2.39
C-5	395	10.4	2.45
C-6	423	11.9	2.51

Table 1. Coating parameters of tested tubes used in present study.

- Data were acquired under steady state within the variation of the wall temperature of 0.1°C in 5 minutes. For each power input, the condenser mass flow rate of liquid was adjusted to maintain the constant pressure. An 8 channel data acquisition module (ADAM-4019) was used to collect the experimental data. The above procedure was repeated for each test surface.

2.1. Fabrication of test surface

In current study, the thermal spraying coating surface was prepared at Metallizing Equipment Co. Pvt. Ltd. (MEC) Jodhpur, India. The oxy- acetylene flame was used to melt copper powder by a spraying gun. For atomization and acceleration of the particles onto heating surface, high pressure air is passed over the molten material to solidify and forming a coating. For supplying the oxygen and acetylene gases, two stage precision regulators were used in this spray system. The spray distance for the coating surfaces is 18 cm. This spray conditions depend on the pressures and flow rates of oxygen and acetylene gases. Oxygen as the oxidizing gas at a pressure of 0.25 MPa and a flow rate of 1.27 m^3/hour with acetylene gas pressure of 0.11 MPa and a flow rate of 1.56 m^3/hour were maintained. The specifications of prepared coated surfaces were as given in **Table 1**. Scanning electron microscope (SEM) images of test surfaces were analyzed using 'Image J' offered by Research Services Branch (RSB) of the National Institute of Mental Health (NIMH). The SEM images of one coated tube of 42 μm thick and plain tube is shown in **Figure 3**. The image analysis procedure is well described by Dewangan

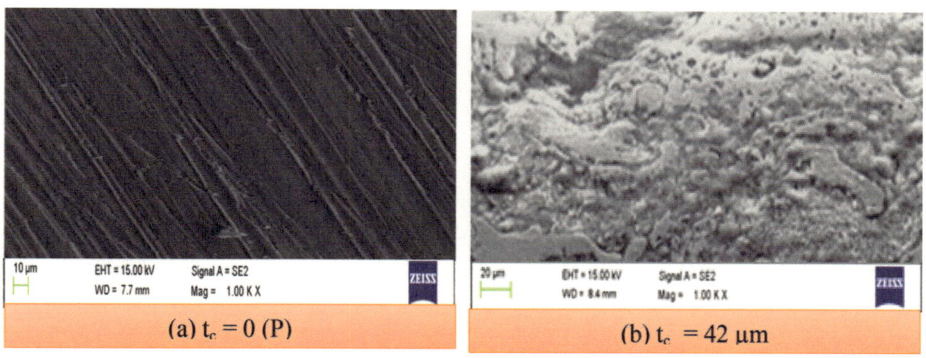

Figure 3. SEM images of plain and coated surfaces.

et al. [18]. This analyze is done to obtain coating parameters of the coated surfaces. As a result, the coating parameters are obtained and as shown in **Table 1**.

3. Results and discussion

In order to establish the integrity of the experimental set-up and verify the temperature measurement in the present test arrangement, preliminary tests have been conducted with R-134a, and R-410A. The heat transfer results of the pool boiling of refrigerants over a horizontal tube are compared with conventional correlations of Stephan and Abdelsalam [19]; Cooper [20]; Cornwell and Houston [21]; Gorenflo [22]; and Jung et al. [23] correlations. The comparisons of the experimental results have been depicted in **Figures 4** and **5**. These figures show the discrepancy between the current experimental data and the predicted results incurred with the same operating condition for different refrigerants. This shows that, the experimental results on pool boiling using the test apparatus and measurement system are reliable. The straight line in each plot signifies that, no discrepancy between the experimental and correlated data.

Dashed lines are appended to show the deviation of calculated values from the experimental data. The experimental result shows ±15% errors for R-134a with mean deviation (MD) ranging from −8.91% to 15.21% and a mean absolute deviation (MAD) of 7.91 to 15.21% for R-134a and ± 19% errors for R-410A with a MD of −19.19 to 18.98% and MAD of 5.85 to 18.98% for R-410A. Among them, the deviations are smallest by using Cornwell-Houston correlation for R-134a. Cooper correlation also shows the minimum deviation the smallest deviation between the experimental and the calculated values for R-134a and R-410A. The variation of imposed

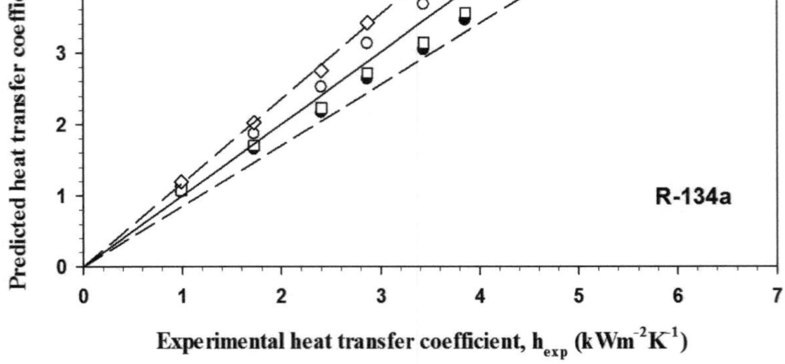

Figure 4. Comparison of experimental data of R-134a with correlations.

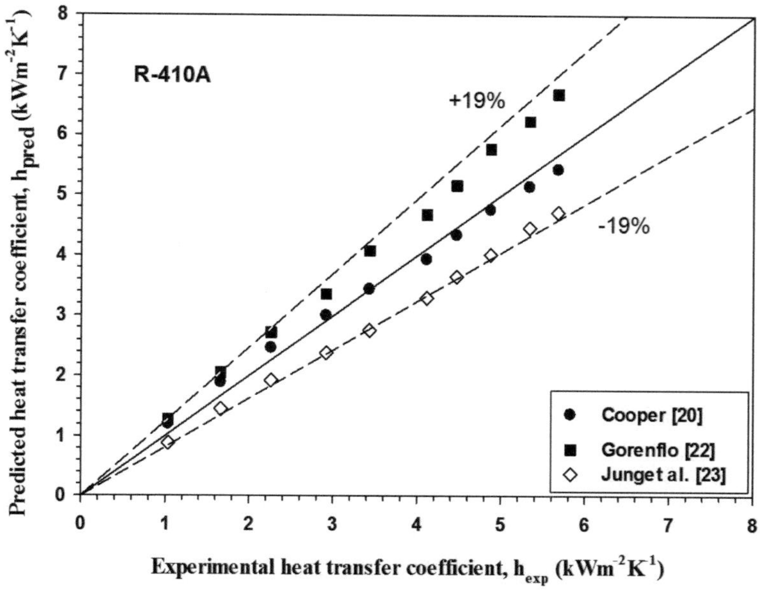

Figure 5. Comparison of experimental data of R-410A with correlations.

heat flux with wall superheat for the boiling of all three refrigerants on a plain heating tube surface at saturation temperature of 10°C. From this observation salient features can be inferred:

- At a given saturation temperature, the wall superheat increases with increase in imposed heat flux and the variation between two by power law, $q \; \alpha \; \Delta T \; n$.
- Heat flux in nucleate boiling of R-134a and R-410A are proportional to the wall superheat raised to a power of 3.86, and 4.22 respectively.

The main reason can also be seen from the physical properties of the tested refrigerants as in **Table 2**.

Figure 6 depicts the variation of heat transfer coefficients of a plain surface for saturated boiling of all three refrigerants at saturation temperature of 10°C with heat flux as a parameter. From this figure, the heat transfer coefficient as a function of superheat for the surface tested and the variation between two can be represented by a power law, $h \propto \Delta T^m$ where m varies between 2.87 and 3.21.

These features can be explained as follows: Wall superheat increases with increase in imposed heat flux on heating tube and therefore variation of the local heat transfer coefficient is found at the top, two sides of the middle and bottom of heating plain tube. This, in turn, increases the wall superheat and thereby value of minimum radius of nucleation sites at which bubble can originate decreases. This can be substantiated by following Eq. (1):

$$r_{cr} = \frac{2\sigma}{\left(\frac{dp}{dT}\right)_{sat} \Delta T_w} \tag{1}$$

Properties	R-134a	R-410A
T_{sat} (°C)	10	10
p (MPa)	0.415	1.09
ρ_l (kgm^{-3})	1261	1128.4
ρ_v (kgm^{-3})	20.23	41.917
k_l (Wm^{-1} K^{-1})	0.088	0.0974
k_v (Wm^{-1} K^{-1})	0.012	0.0132
$\nu_l \times 10^{-4}$ (m^2 s^{-1})	0.0019	0.0013
$\nu_v \times 10^{-4}$ (m^2 s^{-1})	0.0055	0.0030
σ (Nm^{-1})	0.0101	0.0075
M (gmol^{-1})	102.03	72.6
C_{pl} (kJkg^{-1} K^{-1})	1.37	1.58
h_{lv} (kJkg^{-1})	190.74	208.5

Table 2. Properties of tested refrigerants at 10°C [24].

The high heat flux condition enhances the number of nucleation sites and thereby the population of vapor bubbles form on heating surface. The bubbles grow and detach from the heating surface to travel in the pool of refrigerant, which increases the intensity of turbulence near the heating surface and increases heat removal rate. As a result, heat transfer coefficients are obtained to be

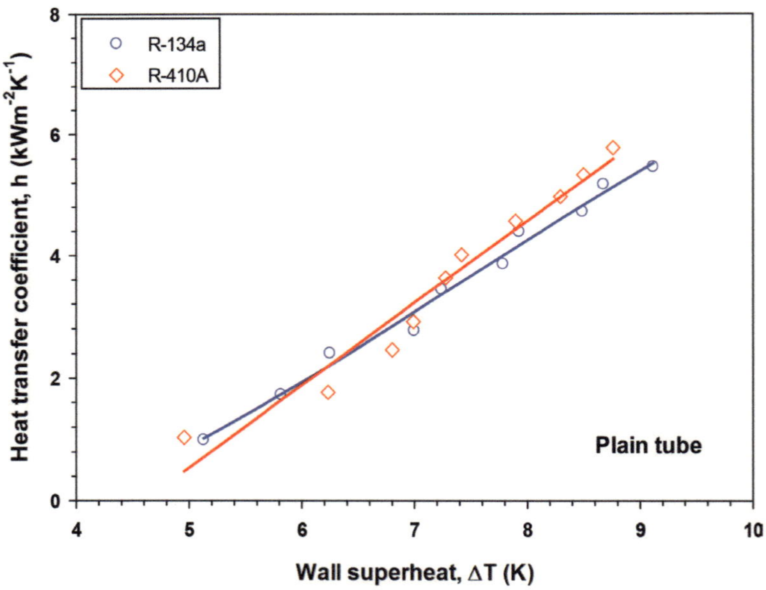

Figure 6. Variation of heat transfer coefficient with wall superheat.

higher at high heat flux condition. The magnitude of heat transfer coefficient at given circumferential position on plain surface is found to differ from liquid to liquid. This is due to variation in thermo-physical properties of liquids under consideration. Similar features have also been observed during pool boiling of R-134a and R-410A refrigerants at saturation temperature of 10°C as shown in **Figure 7** which shows that the local heat transfer coefficients increases with increase in imposed heat flux and shown as $h \propto q^n$ where n values varies between 0.75 and 0.76.

Figures 8 and **9** contain the graphs drawn for heat transfer coefficients versus wall superheat as well as imposed heat flux for seven types of surface. The experimental data points lies in the nucleate pool boiling region as the wall superheat ΔT falls in the range of 2.2–9 K. The coating thickness of 151 μm needs high heat fluxes to achieve a particular wall superheat. Boiling on coated surfaces has differed due to the differences in their surface characteristics. The surface characteristics include the use of different material of substrate, coating material, fabricating method and coating thickness. But in the present study all these surface characteristics namely substrate material, coating material and fabricating method were ensured to be the same. The boiling heat transfer coefficients on the coated surfaces increased with increasing heat flux. The copper coating over heating surface enhances to the formation of porous matrix consisting of micro porous layers. This surface contains large amount of cavities of different sizes. Some of the cavities may have the size that meet the requirement of wall superheat for boiling incipience at low heat flux condition. This causes the coated surface to provide a higher heat transfer coefficient than the plain surface. The very tiny vapor bubbles originate on coated surface due to lower surface tension. Consequently, population of vapor bubble increases and their merging as vapor agglomerates also increases the resistance of bubble departure from coated surface. Therefore, heat removal rate and heat transfer coefficient reduces.

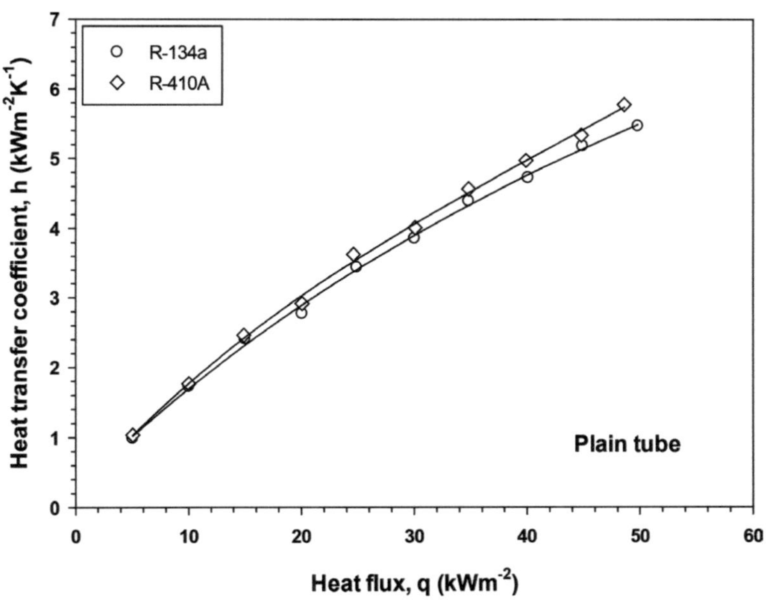

Figure 7. Variation of heat transfer coefficient with heat flux.

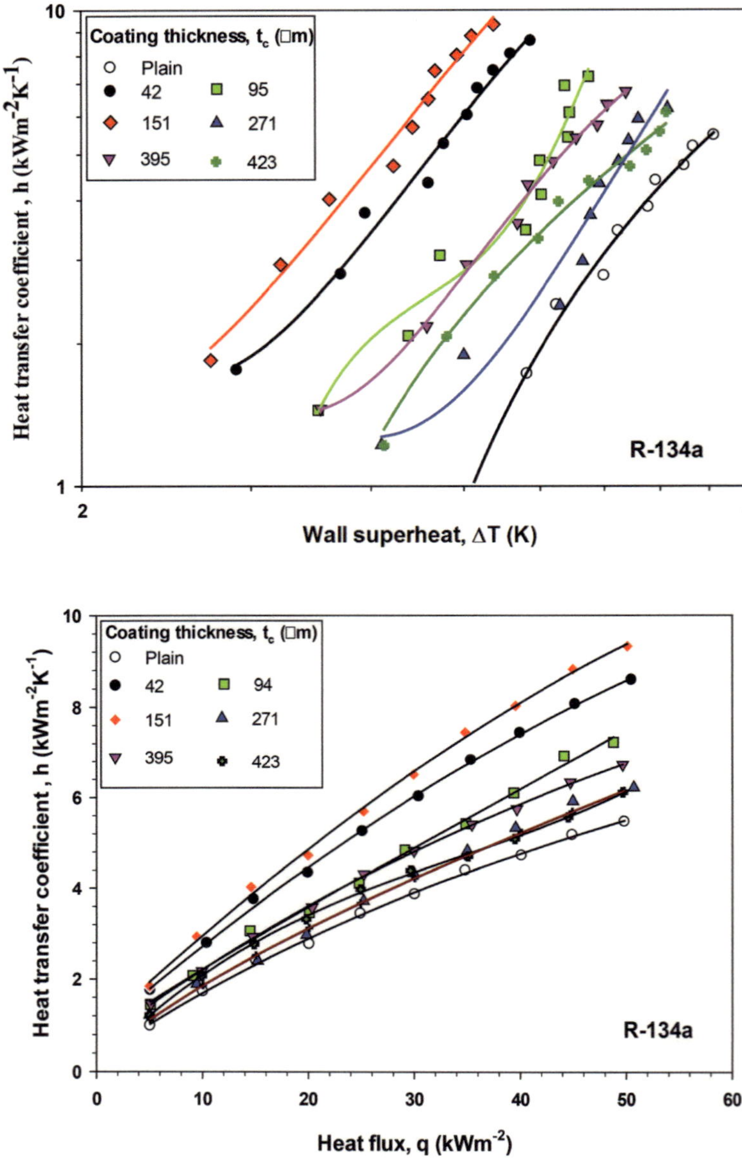

Figure 8. Variation of heat transfer coefficients for R-134a.

As a result, the recirculation of liquid in the inner portion of the coated surfaces increases and it enhances turbulence behavior of bubbles near the heating surface and heat removal rate. The effect of increase in coating thickness reveals two opposing characteristics [25] follows as: heat removal rate increases or decreases as a result of vapor bubble agglomerates and capillary action respectively. Coating thickness and applied heat flux are the main parameters to contribute of each effect. During initial stages of coating, the effect of capillary action is more pronounced than other coating surfaces for enhancing the boiling performance. Therefore,

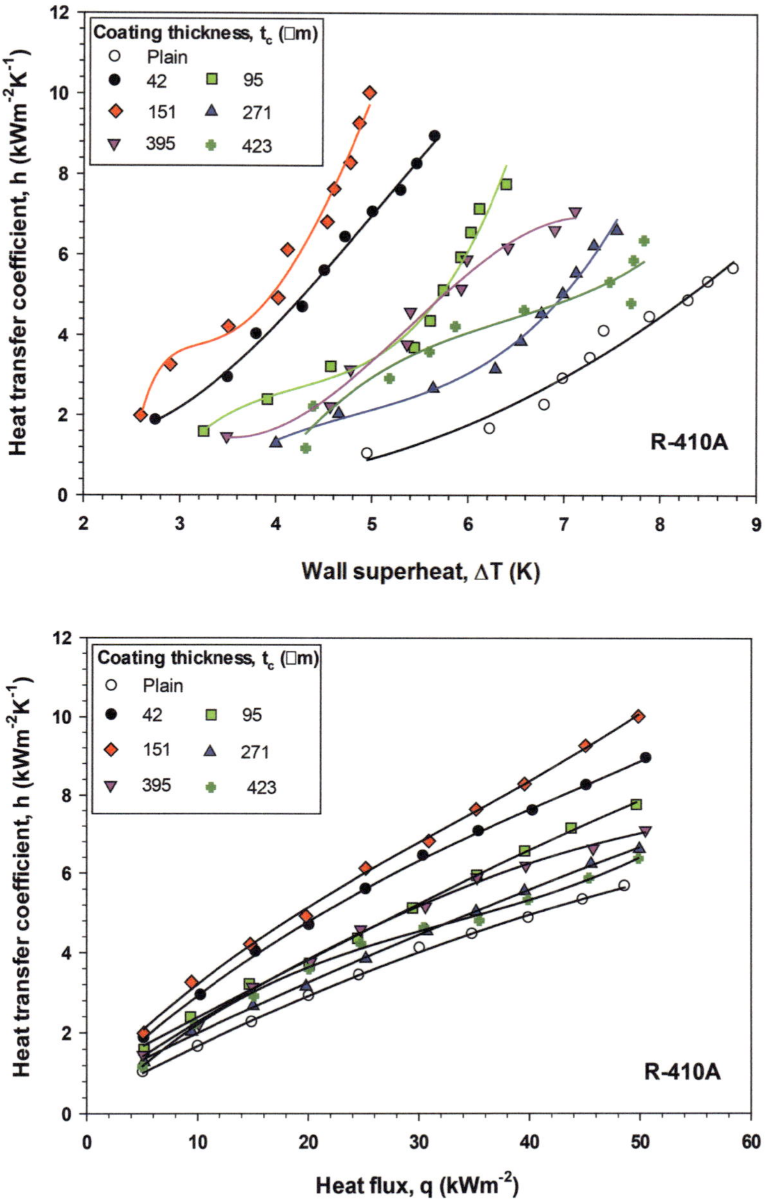

Figure 9. Variation of heat transfer coefficients for R-410A.

the boiling heat transfer coefficient is found to be more on a 42 μm thick coated surface than plain surface.

For an increase in coating thickness from 42 to 95 μm, the above characteristics are observed. However, the nucleation site density is the key parameter to observe the effect of heat removal rate from the heating surface as compared to capillary action. For a given applied heat flux, lower heat transfer coefficient is observed on a 95 μm thick coated tube than 42 μm thick tube due to the combined effect of above characteristics. Further increase in coating thickness from 95 to 151 μm, heat removal rate is more significant than other effects. The reason behind this phenomena is that the rise in recirculation intensity (liquid from bulk rushes to inner layer of the structure with greater intensity). Therefore boiling heat transfer coefficients on 151 μm is found to be greater than that on the 95 μm coated tube. Again increase in thickness of coating from 151 to 271 μm, the effect of heat removal rate is not to be significant than other effects. Thus for increase in coating thickness from 151 to 423 μm, heat removal rate and heat transfer coefficient decreases. For 395 μm coating surface, the intensity of recirculation of liquid play a dominating role than other effects which enhances the heat transfer rate. Due to this phenomenon, higher heat transfer coefficient is achieved on 395 μm coating surface than that on the 271 μm thick coated surface. As can be observed from **Figure 8**, at low heat flux value ≤30 kWm^{-2}, the effect of capillary action is more than that of nucleation site density. This caused the heat removal rate to be higher and therefore heat transfer coefficients on the 423 μm coated tube was found to be higher than that of on the 271 μm coated tube. However, at the heat flux condition ≥ 30 kWm^{-2}, the boiling heat transfer coefficient on 271 μm coating thickness surface is higher than that on 423 μm coating thickness surface. A rise in heat flux increases the number of active nucleation sites to form large number of vapor bubbles on heating surface. In fact, population at some stage becomes so large, and thereby the heat transfer rate and high heat transfer coefficient increases. However, when boiling of refrigerant R-134a occurs on a coated surface, the population of vapor bubble increases due to large heat flux, nucleation sites formed by coating layers.

4. Development of proposed correlation

The evolution of new refrigerants in the market makes the scientist to work for developing a nucleate pool boiling correlation based on a consistent database. Especially, many of the R-22 alternatives azeotropic refrigerant blends and to predict heat transfer coefficients of near azeotropic mixtures, a precise nucleate boiling correlation for pure refrigerants is required immediately. Therefore, a new correlation is developed based upon the present data of three refrigerants following Rudemiller and Lindsay approach [26]. Nucleate pool boiling heat transfer is affected by imposed heat flux, surface specifications, wall superheat, density of liquid and vapor, latent heat of evaporation, characteristic length, other thermophysical properties and their relationship may be summarized in Eq. (2) as described by Incropera [27].

$$h = h\left(q, \Delta T, h_{lv}, \sigma, k, \mu, C_p, \rho, t_c, d_{mp}, \varepsilon\right) \tag{2}$$

The non-dimensional groups can be formed by using the above properties. The major dimensionless groups according to Rudemiller and Lindsay are the Reynolds number *(Re)*, the Jakob number *(Ja)*, the constant heat flux number (N_{cf}) and the geometric scale factor (η). For gas flame coating heating surfaces, the conventional Jacob number (*Ja*) was written in terms of Nusselt number (*Nu*). This correlation considered the effects of system pressure in terms of reduced pressure ratio (p_r), a pressure function $F(p)$ [28] and the refrigerant vapor phase density on boiling heat transfer phenomena and pressure function, $F(p) = 1.8\, p_r^{\,0.17} + 4\, p_r^{\,1.2} + 10\, p_r^{\,10.}$.

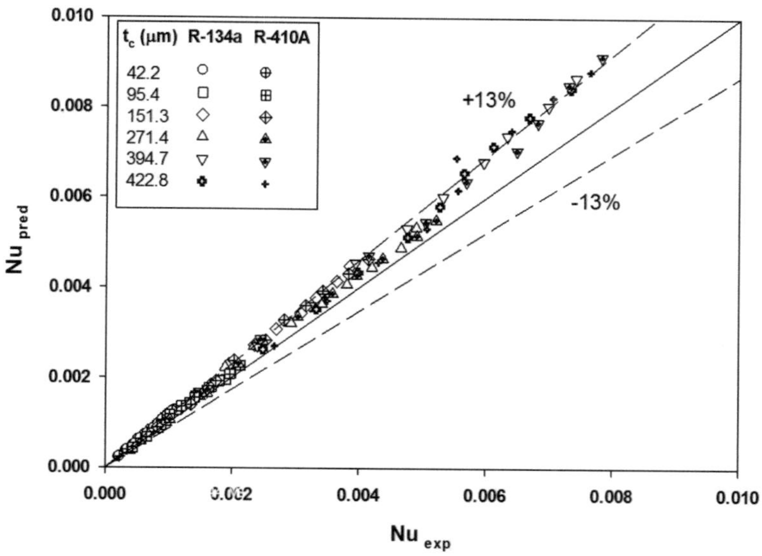

Figure 10. Deviation of present correlation against experimental data.

In this work, the values of constant and exponents depends upon the thermophysical properties of refrigerants R-134a, R-410A and measured experimental data and their values are:

$$C = 0.0132;\ m = 0.312;\ n = -0.043;\ a = 0.852;\ a_1 = 0.429;\ b_1 = 0.612$$

Hence, the relationship can be generated in the generalized form.

$$\begin{gathered} Nu = 0.013\,(Ra)^{0.31}\left(N_{cq}\right)^{-0.04}\left(\frac{1}{\eta_s}\right)^{0.85} F(p)\,(p_r)^{0.43}\left(\frac{\rho_l k_l}{\rho_v k_{eff}}\right)^{0.61} \\ 1.498 \times 10^{-5} \le Ra \le 3.938 \times 10^{-3} \\ 0.00167 \le N_{cq} \le 0.00267 \\ 16.356 \le \frac{t_c}{d_{mp}} \le 113.178 \end{gathered} \tag{3}$$

Figure 10 shows the comparison between the present experimental data with a correlation proposed in this study. The mean deviation was found to be within a maximum error of ±13 percent for all refrigerants tested.

5. Conclusions

In this study, nucleate pool boiling heat transfer coefficients (HTCs) of two refrigerants of R-134a, and R-410A were observed at saturation temperature of 10°C on heating surfaces. Data were collected for the heat flux range from 5 to 50 kW m^{-2} in the interval of 5 kWm^{-2}. As per the experimental results, following conclusions is made:

1. For refrigerant R-134a, the experimental data were predicted by Stephan and Abdelsalam [19], Cooper [20], Cornwell and Houston [21] and Jung et al. [23] correlations within an error band of±15%. For R-410A, the experimental data was predicted within an error band of ±19% by the above correlations. Gorenflo correlation was also predicted the HTCs of R-410A within an error band of±19%. This established the integrity of experimental set-up.

2. At a given saturation temperature and corresponding pressures for all three refrigerants, heat transfer coefficients increase with increase in heat flux. For the same heat flux range, the boiling heat transfer coefficients of R-134a and R-410A are 1.86 and 1.92 times higher than those of plain surfaces respectively.

3. Experimental data for pool boiling of R-134a, and R-410A on copper coated surfaces with different coating thicknesses of copper have been generated for various values of heat flux at saturation temperature of 10°C. Analysis has shown that the heat transfer coefficients increase with increase in imposed heat flux as in the form of $h \propto q\ n$ where n values depend upon the thickness of copper coating and boiling liquid. In fact, the value of n for boiling of refrigerants on coated surfaces is less than those of the plain surfaces.

4. The coated surface of 151 μm thick (C-3) shows the highest enhancement factor of 1.92 among all coated surfaces. In addition, the experimental data for all coated surfaces were correlated in terms of the major dimensionless groups of Nusselt number (Nu), the Rayleigh number (Ra), the geometric scale factor (tc/dp) and the constant heat flux number. The mean deviation was found to be within a maximum error of ±13 percent for all refrigerants tested.

Nomenclature

A	surface area of tube [m^2]
C_{pl}	specific heat of liquid [Jkg^{-1} K^{-1}]
D	tube diameter [mm]
$F(p)$	pressure function
d	diameter [μm]
Gr	grash of number, $Re^2\ P_r^{1/3}$ [–]
h	boiling heat transfer coefficient [Wm^{-2} K^{-1}]
h_{lv}	latent heat of vaporization
I	current [amp]
k	thermal conductivity [Wm^{-1} K^{-1}]
k_{eff}	effective thermal conductivity…, …$\varepsilon k_l + (1 - \varepsilon)\ k_c$ [Wm^{-1} K^{-1}]
M	molecular weight [gmol^{-1}]

Nu	Nusselt number, ht/k_{eff} [−]
p	pressure [MPa]
p_r	reduced pressure, p_{sat}/p_{cr} [−]
Pr	Prandtl number [−]
Q	heat transfer rate [W]
Ncq	constant heat flux number, $\mu_l^2/\rho_l \sigma d_p$ [−]
q	heat flux [Wm^{-2}]
Ra	Rayleigh number, $GrPr$ [−]
Re	Reynolds number, $q\, d_p/\varepsilon\, \mu_l h_{lv}$ [−]
R_s	surface roughness [μm]
t	coating thickness [μm]
T	temperature [K]
ΔT	temperature difference [K]
V	voltage [volt]
Greek letters	
ρ	density [kgm^{-3}]
ν	kinematic viscosity [m^2s^{-1}]
σ	surface tension [Nm^{-1}]
ε	porosity [%]
Subscripts	
c	coating
cq	constant heat flux
cr	critical
exp	experimental
l	liquid
mp	mean pore
$pred$	predicted
s	scale
sat	saturation
v	vapor
w	wall

Author details

Ashok K. Dewangan[1]*, Anil Kumar[2] and Ravi Kumar[2]

*Address all correspondence to: ashokiitr2012@gmail.com

1 School of Mechanical Engineering, Galgotias University Greater Noida, Uttar Pradesh, India

2 Mechanical and Industrial Engineering, IIT Roorkee, Uttarakhand, India

References

[1] Thome JR. Enhanced Boiling Heat Transfer. Washington, DC: Hemisphere; 1990

[2] Webb RL. Principles of Enhanced Heat Transfer. New York: John Wiley & Sons, Inc.; 1994. pp. 311-371

[3] Bergles AE. Techniques to Enhance Heat Transfer, Handbook of Heat Transfer. New York: McGraw-Hill; 1998. pp. 11.1-11.76

[4] Milton RM. Heat Exchange System. U.S. Patent 3; 1968;**384**:154

[5] Milton RM, Gottzmann CF. High efficiency Reboilers and condensers. Chemical Engineering Progress. 1972;**68**(9):56

[6] O'Neill PS, Gottzmann CF, Minton PE. High efficiency heat exchangers. Chemical Engineering Progress. 1973;**69**(7):69-75

[7] Nishikawa K, Ito T, Tanaka K. Enhanced heat transfer by nucleate boiling on a sintered metal layer. Heat Transfer—Japanese Research. 1979;**8**(2):65-81

[8] Hsieh SS, Weng CJ. Nucleate pool boiling from coated surfaces in saturated R-134a and R-407c. International Journal of Heat and Mass Transfer. 1997;**40**(3):519-532

[9] Hsieh SS, Yang TY. Nucleate pool boiling from coated and spirally wrapped tubes in saturated R-134a and R-600a at low and moderate heat flux. Trans ASME: Journal of Heat Transfer. 2001;**123**:257-270

[10] Hsieh SS, Ke CG. Bubble dynamic parameters and pool boiling heat transfer on plasma coated tubes in saturated R-134a and R-600a. Trans ASME: Journal of Heat Transfer. 2002;**124**:704-716

[11] Asano H, Tomita R, Shigehara R, Takenaka N. Heat transfer enhancement in evaporation by thermal spray coating. In: International Symposium on Next-generation Air Conditioning and Refrigeration Technology; Tokyo; Japan. 2010

[12] Scafer D, Tamme R, Steinhagen M, Muller M. Experimental results with novel plasma coated tubes in compact tube bundles. In: Proceedings of Fifth International Conference

on Enhanced; Compact and Ultra-Compact Heat Exchangers: Science; Engineering and Technology; Hobeken, NJ; USA. 2005

[13] Scurlock RG. Enhanced boiling heat transfer surfaces. Cryogenics. 1995;**35**:233-231

[14] Cieslinksi JT. Nucleate pool boiling on porous metallic coatings. Experimental Thermal and Fluid Science. 2002;**25**:557-564

[15] Li Q, Zhang Z, Gao X. Experimental study on pool boiling heat transfer for R22, R407c, and R410A on a horizontal tube bundle with enhanced tubes. Heat Transfer Engineering. 2011;**32**(11-12):943-948

[16] Lakhera VJ, Gupta A, Kumar R. Enhanced boiling outside 8 × 3 plain and coated tube bundles. Heat Transfer Engineering. 2012;**33**(9):828-834

[17] Vasiliev LL, Khrolenok VV, Zhuravlyov AS. Intensification of heat transfer at propane pool boiling on single horizontal tubes. Revue Generale de Thermique. 1998;**37**:962-967

[18] Dewangan AK, Kumar A, Kumar R. Nucleate boiling of pure and quasiazeotropic refrigerants from copper coated surfaces. Applied Thermal Engineering. 2016;**94**:395-403

[19] Stephan K, Abdelsalam M. Heat transfer correlations for natural convection boiling. International Journal of Heat and Mass Transfer. 1980;**23**:73-87

[20] Cooper MG. Heat flow rates in saturated nucleate pool boiling—A wide-ranging examination using reduced properties. Advances in Heat Transfer. 1984;**16**:157-239

[21] Cornwell K, Houston SD. Nucleate pool boiling on horizontal tubes: A convection-based correlation. International Journal of Heat and Mass Transfer. 1994;**37**:303-309

[22] Gorenflo D. VDI-Heat Atlas. Duesseldorf, Germany: VDI-Verlag; 1997

[23] Jung D, Lee H, Bae D, Oho S. Nucleate boiling heat transfer coefficients of flammable refrigerants. International Journal of Refrigeration. 2004;**27**:409-414

[24] Lemmon EW, Huber ML, McLinden MO. NIST thermodynamic and transport properties of refrigerants and refrigerant mixtures—REFPROP version 9.0. 2010

[25] Dewangan AK, Kumar A, Kumar R. Pool boiling of iso-butane and quasi azeotropic refrigerant mixture on coated surfaces. Experimental Thermal and Fluid Science. 2017;**85**:176-188

[26] Rudemiller GR, Lindsay JD. An investigation of boiling heat transfer in fibrous porous media. In: Proceedings of the Ninth Int. Heat transfer Conference. Vol. 5. 1990. pp. 159-164

[27] Incropera FP. Fundamentals of Heat and Mass Transfer. 6th ed. John Wiley & Sons; 2006

[28] Carey VP. Liquid-Vapor Phase Change Phenomena. Washington, DC: Hemisphere; 1992. p. 233

Chapter 7

The Fundamental and Application of Transient Flashing Spray Cooling in Laser Dermatology

Zhi-Fu Zhou and Bin Chen

Additional information is available at the end of the chapter

http://dx.doi.org/10.5772/intechopen.79462

Abstract

Cryogen spray cooling (CSC) has been successfully implemented in laser dermatology such as the treatment of port wine stain. It can protect epidermis from irreversible thermal injuries and increase laser energy, leading to the improvement in therapeutic outcomes. Different from traditional steady spray cooling, CSC is highly transient with short spurt duration (several tens of milliseconds). Besides, CSC can achieve flashing atomization and fine droplets with simple structure nozzles by rapid release of superheat. In this chapter, the mechanism of CSC flashing spray, spray and thermal characteristics of droplets, the measurement method of transient temperature and algorithms for heat flux estimation, and the dynamic surface heat transfer and its relation with spray characteristics are fully discussed. Finally, the heat transfer enhancement of CSC is introduced including alternative cryogens, new nozzles, and hypobaric pressure method to increase the cooling ability, which is essential to improve therapeutic outcome, especially for darkly pigmented human skin.

Keywords: cryogen spray cooling, droplets, dynamic heat transfer, laser dermatology

1. Introduction

When high-pressure liquid is discharged into low-pressure environment below its saturation pressure through a nozzle, the superheated liquid will result in violent flashing boiling spray/flashing spray and two-phase mixture flow [1, 2]. The rapid expansion of vapor bubbles close to the nozzle exit shatters the liquid bulk to produce a finely atomized spray. Different from the hydraulic spray dominated by mechanical breakup, the flashing spray is characterized by explosive atomization, fine droplets, and intense evaporation due to the thermal driving force of unstable superheated liquid returning to its equilibrium state. Low saturation temperature and volatile cryogens, such as R134a, serve as important medium to generate the flashing spray, which has wide applications in many fields due to the fine and uniform droplets with quite low temperature.

As one of the most important applications of flashing spray, cryogen spray cooling (CSC) with R134a has been widely used to protect epidermis from irreversible thermal injuries due to the absorption of laser energy by melanin in the laser treatment of port wine stain (PWS) [3, 4]. PWS is one kind of congenital vascular birthmarks in dermis that occurs in approximately 0.3–0.5% of infants as shown in **Figure 1** [6, 7]. PWS can be associated with significant cosmetic disfigurement and psychological distress. The best choice for the treatment of PWS is laser surgery with selected wavelength to cause permanent thermal damage to the target blood vessels in dermis via the principle of selective photothermolysis [8, 9]. As can be seen from **Figure 2**, however, the competitive absorption of laser energy by melanin in normal tissue (especially in epidermis) will not only reduce the therapeutic effect but also cause irreversible thermal damage to epidermis, due to the close absorption peak of laser energy at the selected wavelength (typically 585 and 595 nm) between the oxyhemoglobin (HbO_2) and melanin. In CSC, a pulsed cryogen spurt with a duration of several decades of milliseconds (no more than 100 ms) is applied to the skin surface prior to laser irradiation in order to selectively cool down the epidermis, aiming to achieve the largest possible temperature difference between epidermis and deeper malformed blood vessels [10–12]. Consequently, CSC enhances the threshold of laser radiant exposure for irreversible thermal damage to epidermis, and higher laser energy can be employed in laser surgery, resulting in the improvement of therapeutic outcomes.

Although CSC-assisted laser therapy has been regarded as the gold standard of PWS treatment, lots of clinical studies have demonstrated that majority of PWS failed to clear completely (less than 20%) [7]. One of the main reasons for the failure of PWS complete clearance is attributed to the insufficient cooling capacity of CSC, especially for the darkly pigmented human skin, which limits the laser energy in clinical surgeries. In general, treatment efficacy correlates negatively with increased melanin content in epidermis due to the strong absorption of laser energy [6]. Therefore, the enhancement of cooling efficiency is quite essential for the further improvement of PWS therapy. Cooling efficiency is usually characterized by

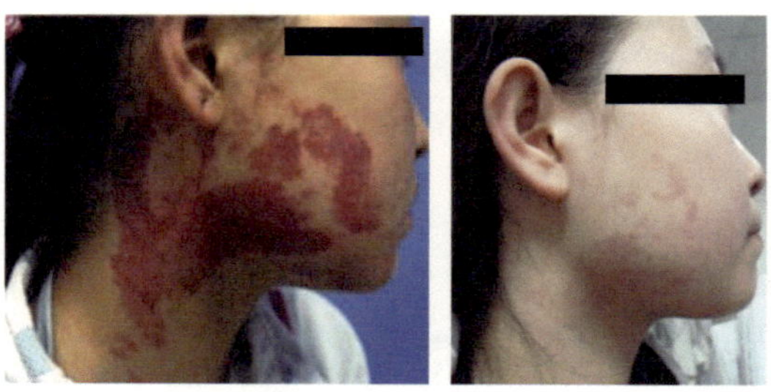

Figure 1. PWS before and after laser treatment (Curtsey of Drs. Wang and Yin at Laser Cosmetic Centre of 2nd hospital of Xi'an Jiaotong University, Xi'an, China) [5].

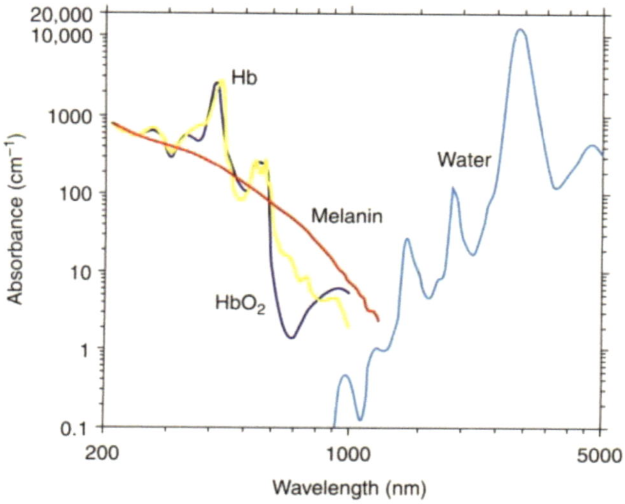

Figure 2. Absorption spectra of major chromophores in skin [9].

dynamic heat transfer, i.e., surface temperature and heat flux on cooling substrate, which greatly depends on spray characteristics including droplet diameter, velocity, and temperature. Therefore, the spray characteristics and heat transfer dynamics of CSC have been studied extensively through experimental method and numerical simulation during the past 20 years.

In this chapter, we will first present a brief introduction of mechanism of flashing spray, and then, we discuss the spray and thermal characteristics of droplets in flashing spray of CSC. Next, we will explain measurement methods of transient surface temperature and algorithms to estimate surface heat flux induced by CSC. The dynamic heat transfer and its qualitative relation with spray characteristics of CSC will also be discussed. Finally, we introduce several heat transfer enhancement methods to increase cooling capacity of CSC including nozzle design, alternative cryogens, and hypobaric pressure method.

2. Flashing spray mechanisms

Usually, flashing spray can be obtained by isothermal pressure drop (line OA in **Figure 3**), during which the liquid from its initial stable sub-cooled condition (O) quickly goes into a metastable superheated state (A). The faster the process, the more violent flashing spray can be reached. The spinodal line close to the liquid saturation line represents the thermodynamic limit of metastable liquid region, which can be evaluated by $(\partial P/\partial v)_T = 0$. The extent of metastable liquid is usually represented by superheat degree in terms of $\Delta T = T1 - T_{sat,\,PA}$ or $\Delta P = P_{sat,T1} - P_A$. The superheated liquid always has the nature of returning to its equilibrium state through rapid phase transition, which starts from vapor bubble nucleation through bubble growth to the eventual stable liquid-vapor equilibrium. There are two types of nucleation processes identified as homogeneous and heterogeneous nucleation. The former

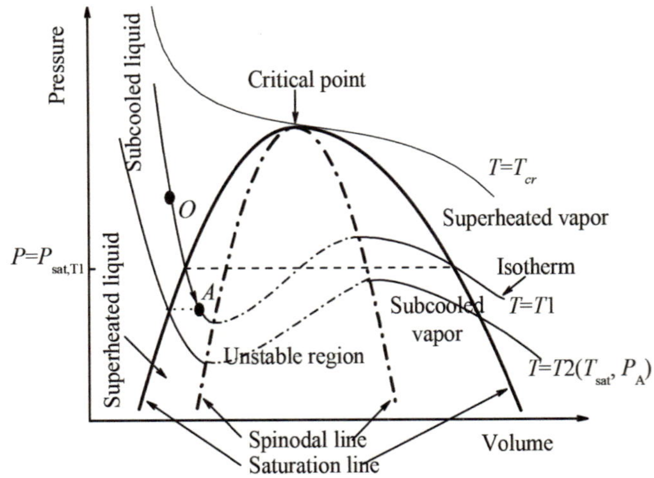

Figure 3. Schematic of pressure-volume isotherm for a pure fluid [13].

is a fundamental mechanism of first-order phase transitions, occurring in the volume of the superheated pure liquid, whereas the latter initiates from the boundaries of the liquid phase with either impurities (including dissolved gas micro-bubbles and dust) or container walls and is a second-order phase transition. Once the nuclei are larger than a critical value, they will grow further to generate bubbles.

According to the location of phase transition happening, there are two models of flashing spray [14]. One is the external flashing, and phase change only occurs outside the nozzle, which is

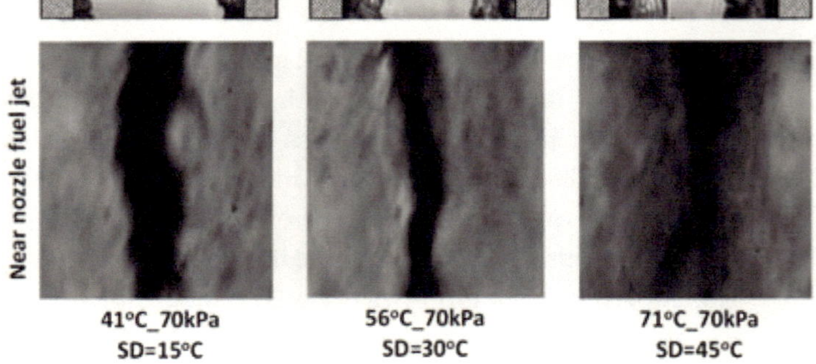

Figure 4. Effect of fuel temperature on internal flow and near nozzle n-pentane fuel spray characteristics [17].

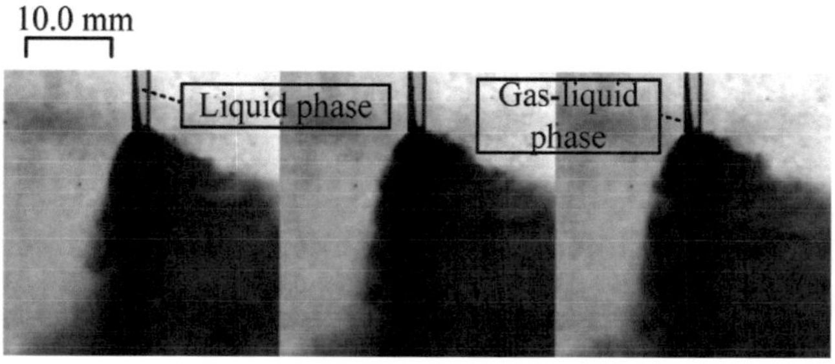

Figure 5. Internal flow and spray pattern near nozzle field of R134a flashing spray [18].

usually difficult to achieve. The other is the internal flashing and more commonly occurs in various applications in which phase change takes place either within the internal nozzle or simultaneously inside and outside the nozzle. In this model, different spray pattern of two-phase flow before discharge can be obtained from the initial bubbly flow to slug flow and annular flow with increasing the superheat degree [14–16]. It is noticed that the internal flow governs the atomization behavior outside the nozzle. For such case, more violent atomization and larger spray angle occur with higher superheat degree, i.e., higher void fraction. Usually, high superheat and long nozzle with small diameter are more likely to generate internal flashing spray, while low superheat and short nozzle with large diameter have more possibilities of inducing external flashing spray. Most previous studies about internal flashing spray using water and hydrocarbon mediums indicate that bubble formation inside nozzle is in favor of atomization with larger spray angle as shown in **Figure 4** [17]. However, recent study of R134a flashing spray shows the opposite trend that phase change/bubble formation inside nozzle mitigates the intensity of flashing atomization outsize the nozzle with smaller spray angle shown in **Figure 5** [18]. This is explained by the fact that the prior phase change inside the nozzle greatly decreases liquid/vapor temperature, corresponding to a lower superheat degree as they are discharged.

3. Spray and thermal characteristics of flashing spray

The release of superheat of flashing spray is quite rapid and only occurs near nozzle field, resulting in the breakup of liquid into fine droplets. Then droplets experience equilibrium evaporation as well as heat and momentum interaction with ambient gases. As a result, their diameter, velocity, and temperature continuously vary within the flow field. These spray and thermal characteristics have an important effect on heat transfer performance on cooling substrate. Currently, R134a is the only cryogen that gets the allowance of application in laser surgery by US Food and Drug Administration (FDA). Therefore, most attentions have been paid on R134a spray and thermal characteristics, and stainless steel straight tube nozzle is usually employed with length ranging in 20–80 mm and diameter ranging in 0.3–1.5 mm. Droplet diameter and velocity can be simultaneously measured by Phase Doppler Particle Analyzer (PDPA) based on the principle of light interface, or droplet diameter can be also measured by Malvern based on Mie scattering theory. Both of the measurement methods are noninvasive, having no disturbance to spray field.

Pikkula et al. [19, 20] investigated the effect of droplet velocity and diameter on heat removal during R134a flashing spray, employing four different types of nozzles including full and hollow cone spray nozzles and straight tube nozzle. They found that nozzle type greatly influences droplet diameter and mass flow rate while has less effect on heat removal on cooling substrate. They also found a nonlinear relationship between heat removal and Weber number, and a higher Weber number was in favor of cooling capacity. Karapetian et al. [21] explored mass flow rate and droplet velocity on surface heat flux of R134a flashing spray using eight straight-tube nozzles with four different diameters (0.57, 0.83, 1.08, and 1.33 mm) and two different lengths (8 and 65 mm). Their results showed that change in mass flow rate on heat flux was more important than that of droplet velocity. For fully atomized sprays, however, larger droplet velocity could substantially enhance surface heat flux. Aguilar et al. [10, 22, 23] also investigated the effect of nozzle diameter and length on spray characteristics of R134a flashing spray. In contrast to nozzle length, nozzle diameter had greater influence on the spray characteristics. Larger diameter nozzle with shorter length generated jet-like spray with higher droplet velocity and larger droplet diameter, while smaller diameter nozzle caused more finely atomized spray with lower droplet velocity and smaller droplet diameter. Vu et al. [24] further investigated the effect of nozzle length (20, 40 and 80 mm, with same inner diameter of 0.5 mm) on spray behavior of R134a flashing spray, but no obvious difference of external spray characteristics was observed. They also proposed a simple two-phase flow model inside straight-tube nozzle to explain the acceleration of droplet velocity near nozzle field that the higher velocity airflow accelerated the lower velocity droplet. Zhou et al. [25] obtained droplet density, diameter, and velocity evolution along both radial and axial directions within flow field. A sharp decrease in droplet diameter and an increase in droplet velocity near nozzle field were found, while their variation became more gradual in the far field of spray. They proposed nondimensional correlation of droplet density and velocity distributions based on the experimental data. Zhou et al. [26] also investigated the effect of different cryogens R134a, R407C, and R404A on spray characteristics. Their study indicated that more volatile cryogen induces better atomization, generating higher droplet velocity and smaller size diameter. Yildiz et al. [27–29] studied spray characteristics of R134a flashing spray with larger inner nozzle diameter (1–5 mm) and subcooled cryogen liquid pressurized in a high-pressure container. Much different with the atomization phenomenon of previous studies by Pikkula, Aguilar, Vu and Zhou, they observed that flashing only occurred outside nozzle exit that liquid column suddenly shattered at certain distance from nozzle tip, while flashing already happened inside nozzle for other studies. This kind of external flashing spray generated far larger droplet diameter than those by internal flashing spray.

Compared to the investigation on droplet diameter and velocity, there were only few studied on droplet temperature. A big challenge is the difficulty of measuring low droplet temperature in volatile flashing spray through noninvasive methods. Alternatively, the intrusive method with a fine thermocouple is usually employed to obtain droplet temperature quantitatively, while disturbance cannot be avoided to spray. Aguilar et al. [10] and Yildiz et al. [30, 31] investigated droplet temperature evolution along central spray axis with the intrusive thermocouple. Droplet temperature first experiences a fast exponential decay near nozzle field and then decreases more slowly with spray axial distance until reaching the minimum temperature. Although nozzle diameter affects the decreasing rate of droplet temperature, the minimum temperature of R134a flashing spray is independent of nozzle diameter, always equaling to −60°C. The distance of the minimum temperature occurring was defined "minimum temperature distance" by Polanco et al. [32]. They also reported a dimensionless minimum temperature distance ranging from 150 to 170 (normalized by nozzle diameter) for R134a and propane flashing spray. Later, Zhou et al. [25] conducted a much more comprehensive study on the global thermal characteristics of R134a flashing spray with an inserted thermocouple of 0.1 mm. It was found that a warm core existed near nozzle field where droplet radial distribution showed "W" shape, while it transited to a "U" shape in the far field of spray as shown in **Figure 6**. They also proposed two thermal widths (thermal width R_{STW} and cold width R_{SCW}) and a correlation to describe droplet temperature variation along axial distance.

In addition to the experimental studies, numerical simulation provides alternative way to study flashing spray. However, most of numerical studies concern the fuel flashing spray in internal combustion engines. Few numerical studies pay attention to cryogen flashing spray with very low saturation temperature and high volatility. Recently, Zhou et al. [33] explored the spray morphology, droplet diameter, velocity and temperature behavior of R404A flashing spray through OpenFOAM, and their result matched experimental data reasonably although some discrepancy existed in the near nozzle field. Wang et al. [34] conducted numerical study on R134a flashing spray near the nozzle exit by a three-dimensional vortex method and explored the effect of spray back pressure, ambient temperature, and mass flux on the heat flux on cooling surface. Their results could predict spray distance and back pressure for specific heat flux peak at any ambient temperature and nozzle diameter, which provided personalized and precise reference for parameter selection in clinical spray cooling.

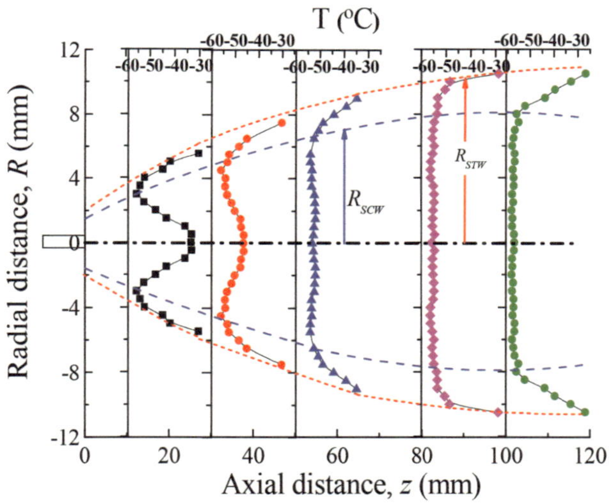

Figure 6. Radial temperature distribution of R134a flashing spray at various axial distance [25].

4. Dynamic surface heat transfer of CSC

4.1. Transient surface temperature measurement and heat flux calculation

During the short pulse spray less than 100 ms, surface temperature and heat flux undergo extremely rapid variations as droplets impinge on cooling substrate. Thus, surface temperature measurement and heat flux calculation are challenges and are also important for CSC. Additionally, it is quite difficult to carry on experiments with real human skin. Usually, researches use epoxy resin and plexiglass as skin phantoms in experiments due to their similar thermal physical properties with skin tissue. There are two methods to measure the transient surface temperature as shown in **Figure 7**. One is the indirect method that commercial thermocouples of round or plate-shaped joints with at least 50 μm coverer by a thin layer of aluminum foil are placed on the cooling substrate surface. As a result, the thermocouple is not directly exposed to the droplets of the spray, which usually gives a poor response of the surface temperature in the order of a few milliseconds [36]. The other one is using thin film thermocouple (TFTC) with thickness of ~2 μm to measure the surface temperature directly, since TFTC is deposited onto the cooling substrate surface through Magnetron technique. It has perfect contact with the underlying substrate and a fast response time (1 μs) [4]. Therefore, the surface temperature measured by TFTC varies much faster than by indirect method as shown in **Figure 8**.

It is far more difficult to measure the time-varying heat flux directly at the solid surface than to measure surface temperature. Alternatively, it is usually estimated from the temperature measurement made at accessible locations, which is termed an inverse heat conduction problem (IHCP). Several analytical and numerical methods have been proposed for the solution of IHCP, such as specified sequential function method (SFSM), regularization method, and transfer function method, among which the most popular method for estimation of surface heat flux is SFSM. Tunnell et al. [19, 37–39] predicted surface heat flux during and following cryogen spurt using the SFS method based on internal temperature measurement, in which a wire-like thermocouple with a bead diameter of 30 μm was imbedded in epoxy resin. Aguilar et al. [36, 40, 41] also employed the SFS method to estimate surface heat flux during CSC

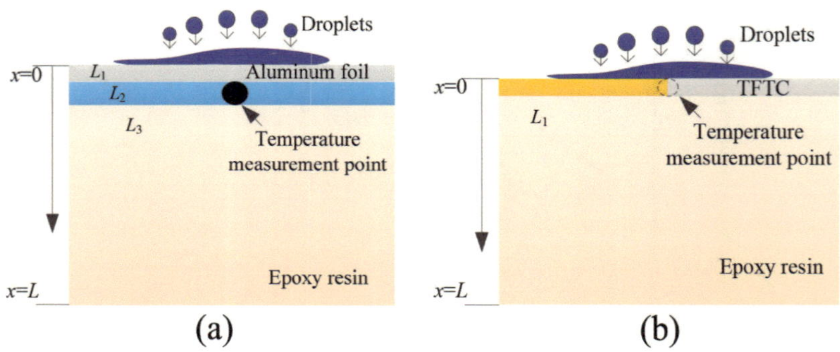

Figure 7. Schematic of the two different surface temperature strategies [35]. (a) Indirect measurement of surface temperature by covered thermocouple; (b) Direct measurement of surface temperature by TFTC.

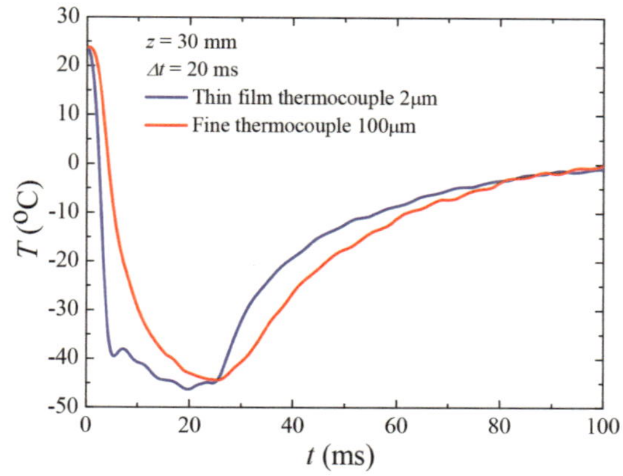

Figure 8. Comparison of transient surface temperature measured by indirect and direct methods [35].

based on internal temperature data measured by indirect measurement method. Using the same temperature measurement method, Franco et al. [42, 43] calculated the surface heat flux by solving a direct problem through Duhamel's theorem, where the measured temperature was treated as the real surface temperature. Zhou et al. [4, 44, 45] also predicted surface heat flux with Duhamel's theorem but based on surface temperature measured directly by TFTC. It should be noticed that the Duhamel's theorem is only applicable to the calculation of surface heat flux at the condition of surface temperature measured directly. Otherwise, it causes much error to the heat flux result. Later, Zhou et al. [35] developed a new method based on Duhamel's theorem that could predict surface heat flux accurately in both cases of indirect and direct surface temperature measurement methods. However, all the previous studies about CSC, heat transfer within the cooling substrate during CSC was treated as one dimensional heat conduction problem, only taking into account heat transfer along the depth direction while ignoring the latent heat transfer. Recently, [46] proposed a two dimensional model to predict surface heat flux through a filter solution method. Their result indicated that the maximum heat flux was 13.6% higher than that predicted by 1D heat conduction model.

4.2. Parametric study on heat transfer of CSC

Great efforts have been made to investigate the heat transfer characteristics of R134a spray cooling under various conditions during the past two decades. Aguilar et al. [36, 47] extensively studied the effect of spurt duration and spray distance. Their results showed that the surface temperature and heat flux were significantly affected by the spray distance while little influenced by the spurt duration. They also studied the effect of the spray angle between the nozzle and the cooling surface [41]. The result indicated that there was insignificant effect on heat flux and heat extraction at a wide range of spray angles (15 ~ 90°). Jia et al. [40] investigated the initial temperature of the cooling substrate. It was found that a higher initial temperature could cause a greater heat flux. Franco et al. [43] and Wang et al. [44] investigated the radial variation of heat transfer during CSC and found a sub-region of uniform cooling at the center of the sprayed surface. Pikkula et al. [19] explored heat removal during CSC using different nozzles and found no apparent variation, despite the relatively large difference in cryogen mass output. Majaron et al. [48] observed water condensation and frost formation on cooling substrate during CSC. It was found that the latent heat deposited by condensation of water vapor and subsequent frost formation significantly impair CSC cooling rate while may reduce the risk of cryo-injury associated with prolonger cooling. Ramirez-San-Juan et al. [49] investigated the effect of ambient humidity on light transmittance during CSC due to scattering of light by the spray droplets and subsequent water condensation/freezing on skin surface. They found that light transmittance was greatly reduced with increasing humidity caused by more intense condensation of water vapor at higher ambient humidity. Vu et al. [50] used a wire meshes in the spray path to enhance cryogen spray atomization in order to enhance cooling efficiency on cooling substrate. However, this passive way did not succeed in enhancing cooling efficiency, while it could prolong cooling duration for larger nozzle. Majaron et al. [51] employed a split injection strategy (intermittent spray) to eliminate liquid film formation on cooling substrate of metal disk, which impaired heat transfer due to relatively low thermal conductivity, and found that highest cooling rates reached at moderate duty cycle levels. Basingerd et al. [52] found that skin indentation due to the force of an impinging cryogen spray reduced heat transfer efficiency of CSC because of cryogen accumulation compared to the flat surfaces. But once indentation existed, larger indentations produced a higher maximum heat flux caused by the stronger convective flow within the cryogen pool.

Heat transfer coefficient served as a boundary condition is quite crucial for calculation of temperature distribution within cooling substrate or skin during CSC. However, previous studies did not yield a general correlation of heat transfer coefficient of CSC. Recently, Tian et al. [53] found that transient cooling could be divided into two stages, namely, fast boiling cooling and film evaporation cooling. The similarity of dynamic heat flux with different cryogens, nozzles, and substrates was observed, and a dimensionless correlation was proposed, as presented in **Figure 9**. As to the maximum heat flux, a nondimensional correlation was proposed by coupling the Jakob number (Ja), Reynolds number (Re_l), and Weber Number (We), as shown in **Figure 10**.

Although CSC selectively cools and protects epidermis from thermal injuries in dermatologic laser surgery, aggressive cooling may have the risk of causing cryo-injuries to epidermis and dermis. Kao et al. [54, 55] studied the potential for epidermal and dermal injury exposed to continuous CSC spurt durations of 10, 20, 40, 80, 100, 200, and 500 milliseconds through experimental study using a vitro model of human skin (RAFT). It was found that spurt duration longer than 100 ms could result in epidermal injury acutely, and 500 ms spurt duration of CSC even could cause decreased fibroblast proliferation to dermis. Li et al. [56, 57] also simulated the cooling process and evaluated the potential cold injury of CSC with R134a, R407C, and R404A through a multiscale model, in which cold injury was recognized once cells were dehydrated or the ice formed intracellularly. They reported spurt duration causing cold injury were 3.3, 2.2, and 1.9 s for R134a, R407C, and R404A, respectively, which was much longer than that reported by Kao et al. [54, 55]. For actual clinical application of CSC, the spurt duration is controlled within 100 ms, which should not cause cold injury to both epidermis and dermis.

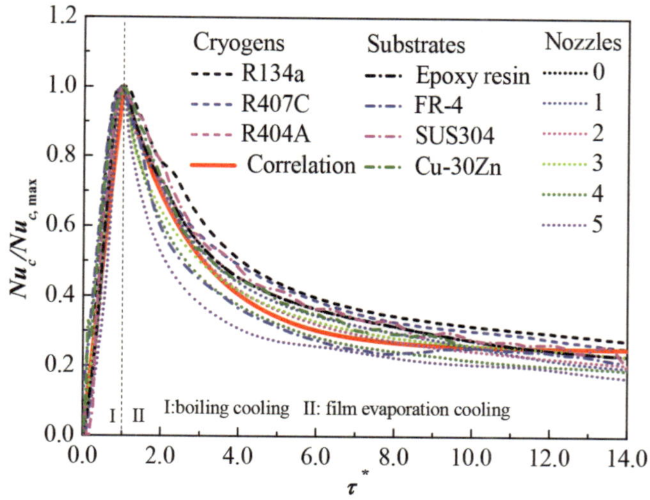

Figure 9. Similarity of nondimensional heat flux at various conditions, different cryogens, cooling substrates and nozzles.

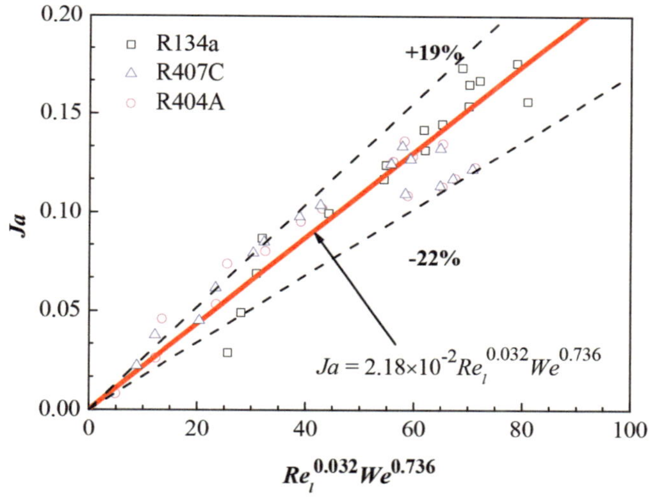

Figure 10. Dimensionless correlation of maximum heat flux [53].

5. Heat transfer enhancement of CSC

CSC is of much importance during dermatologic laser surgery, reducing thermal injury to epidermis, and increasing laser energy for treatment of PWS. However, the therapeutic outcome is still not satisfied (the complete clearance of PWS is less than 20%), and thermal injury of epidermis commonly occurs especially for dark human skin like Asians, which greatly limits the allowance of laser energy in clinical surgery [7]. One of the most important reasons lies in the insufficient cooling efficiency induced by CSC. Thus, several strategies have been proposed to further improve the cooling efficiency in past 20 years.

New cryogens with a lower saturation temperature and higher volatility are essential for enhancing the cooling capacity of CSC based on the current spray technique. Alternatively, cryogens R407C and R404A meet the requirement, with a far lower saturation temperature of −43.6 and −46.5°C at 1 atm, respectively. Moreover, they share a similar feature with R134a in that they are nontoxic to the human body and friendly toward ozone depletion. Dai et al. [58] first carried out preliminary studies on the heat transfer of R404A spray cooling for the laser dermatology of PWS. They used a 30-μm diameter thermocouple, embedded at a 100-μm depth below the phantom surface, to measure the temperature at a long spray distance (85 mm). Their work revealed that R404A spray cooling produced a lower surface temperature than that of R134a and did not cause any cold injuries to skin within 300-ms spray duration. Zhou et al. [45] investigated the effect of spray distance and spurt duration on heat transfer characteristics during CSC with R404A through a thin film thermocouple of 2-μm thickness. It was found that the minimum surface temperature appeared at a spray distance of 30 mm, and a slightly higher heat flux could be obtained for a shorter spray distance, whereas the spurt duration had little effect. Later, Zhou et al. [26] did a comparative study on the spray behavior and heat transfer dynamics using different volatile cryogens (R134a, R407C, and R404A), further demonstrating the priority of R404A spray cooling with lowest surface temperature and highest heat flux with regard to the cooling capacity. The maximum heat flux was 20% higher by R404A than that by R134a, while the liquid film resistance time was 50% less than that by R134a spray cooling. In addition, quite different from the divergent morphology of R134a spray, R404A presents a convergent spray pattern toward axial distance with much smaller spray width after its rapid expansion at nozzle exit as shown in **Figure 11** because of its higher volatility, which is helpful to the precise control of cooling region in clinical surgery [59]. All the above studies indicate that R404A has much potential of substituting the current R134a in the application of laser dermatology, especially in the case of darkly pigmented human skin. However, all R134a, R407C, and R404A have a high global warming potential (GWP > 1000) which limits their use in future according to the Kigali

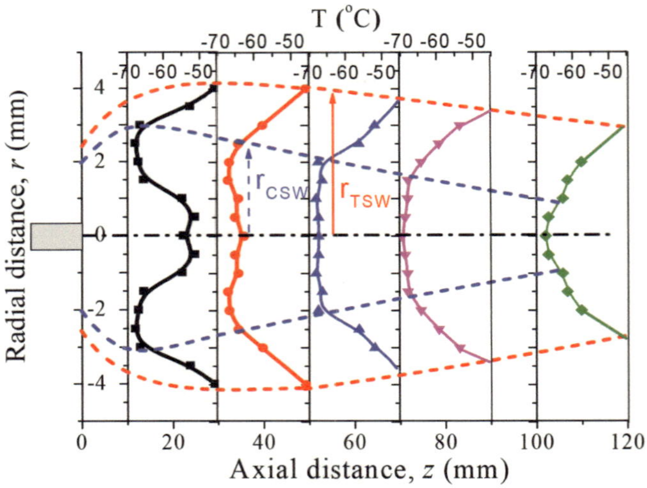

Figure 11. Radial temperature distribution of R404a flashing spray at various axial distance [59].

agreement signed in 2016. Wang et al. [60] firstly suggested using R1234yf as a substitute of traditional HFCs in CSC application due to its extremely low GWP of 4 and its similarity of thermal property with R134a. Their experimental results indicated that the cooling capacity of R1234yf spray was a bit lower than that by R134a, but its cooling efficiency could be enhanced through reducing the superheat degree of flashing spray.

Aguilar et al. [61, 62] proposed a hypobaric pressure-modulatable technique, in which a close chamber and a vacuum pump were used to lower the spray back pressure, by which the surface temperature was reduced and the heat extraction from the cooling substrate was enhanced. Moreover, low pressure could induce vasodilation, which facilitated the easier absorption of laser energy by small blood vessels [63]. This technique was recommended as a promising mechanical way of offering better protection for the epidermis and further improving the therapeutic outcome of PWS [6]. Recently, Zhou et al. [4, 64] employed this technique to investigate the coupling effect of spray distance and back pressure on surface temperature and heat flux during CSC with R134a and R404A. The results proved that the droplet diameter and velocity were greatly reduced due to the enhanced evaporation rate through lowering the back pressure. Heat transfer presented different varying trend with spray back pressure at different spray distances. For R134a spray cooling, the maximum heat flux could be enhanced from 247 to 641 kW/m^2, when the back pressure was lowered from atmospheric pressure to 0.1 kPa at very short distance of 10 mm as shown in **Figure 12**. However, aggressively lowering back pressure below the transitional pressure could lead to a reduction of maximum heat flux at other longer spray distances. Therefore, the spray distance and back pressure should be appropriately deployed to ensure a high heat flux. It is recommended to use the shorter spray distance (10 ~ 20 mm), when the pressure is no larger than the transitional pressure, whereas to use the longer spray distance (40 mm), when the pressure is around 50 kPa. For the atmospheric pressure, 30 mm is the most suitable spray distance.

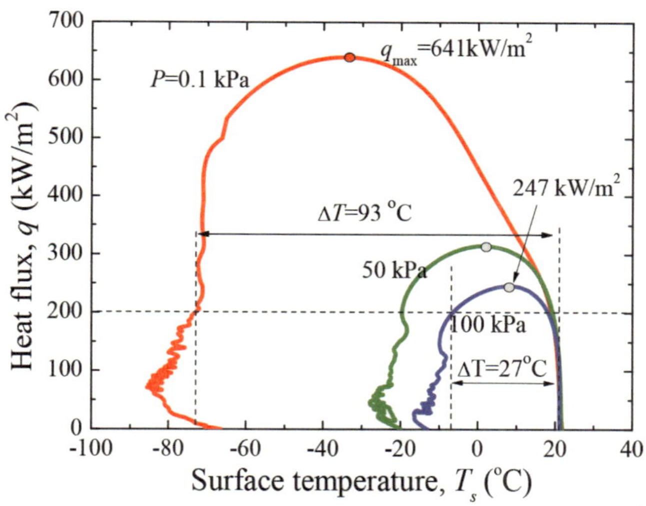

Figure 12. Variation of surface heat flux with surface temperature under different spray back pressures at the distance of 10 mm [4].

Wang et al. [44, 65] designed a cylindrical expansion chamber nozzle with different aspect ratios of chamber length to width for CSC application. Cryogen liquid experienced a first flashing atomization to release part of superheat inside the expansion chamber before discharge from the subsequent straight-tube nozzle. Therefore, this kind of nozzle resulted in much lower droplet temperature and narrower spray width near nozzle field, which enhanced the maximum heat flux by 60% compared to the traditionally used straight-tube nozzle at the optimized aspect ratio of ~0.5.

6. Conclusions

With the rapid development of laser dermatology, protection of normal skin tissue becomes increasingly important for improvement of therapeutic outcomes and attracts more and more attentions in order to optimize the cooling efficiency. This chapter presents a brief review of the progress of cryogen spray cooling in laser surgery for the treatment of port wine stain, focusing on flashing spray mechanism, spray and thermal characteristics of droplets, dynamic heat transfer, and strategies of heat transfer enhancement. Using simple structure nozzle can achieve good atomization characterized by fine droplets with diameter from 5 to 20 μm and velocity usually below 60 m/s induced by internal flashing and release of remaining superheat of liquid at nozzle exit. The explosive atomization near nozzle field also leads to low droplet temperature below its saturation temperature, and the continuous evaporation in flight further reduces droplet temperature. The impingement of droplets results in a strong dynamic heat transfer involved heat conduction, impinging convection, nucleate boiling, and surface evaporation on cooling surface during the transient flashing spray, which rapidly lowers surface temperature and removes heat from epidermis while almost has no effect on the temperature of underlying layer of dermis. Several new methods of using expansion chamber nozzle, substitute cryogens and hypobaric pressure method have been proposed to enhance its cooling efficiency.

Although great progresses have been achieved in both clinic practice and physical understanding of cryogen spray cooling in the past two decades, many issues still remain. First, the fundamental and governing principal of internal flow inside the mini-nozzle on the external flashing spray is not clear enough and needs further exploration. Second, noninvasive measurement of droplet information for dense spray near nozzle field is still a big challenge especially for low temperature measurement of two phase flow. Third, although heat transfer performance greatly depends on spray and thermal characteristics, the fundamental of their relationship is quite complicated and affected by various factors in this transient flashing spray. It is important to develop a theoretical model to connect these two physical processes and thus can have a better guidance of cooling optimization. Last but not the least, although several strategies of heat transfer enhancements have been recommended from the mechanical perspective, much effort of clinical study is needed to demonstrate the feasibility and priority of these new methods in clinical practices.

Author details

Zhi-Fu Zhou and Bin Chen*

*Address all correspondence to: chenbin@xjtu.edu.cn

State Key Laboratory of Multiphase Flow in Power Engineering, Xi'an Jiaotong University, Xi'an, China

References

[1] Günther A, Wirth K-E. Evaporation phenomena in superheated atomization and its impact on the generated spray. International Journal of Heat and Mass Transfer. 2013;**64**:952-965

[2] Cheng W-L, Chen H, Hu L, Zhang W-W. Effect of droplet flash evaporation on vacuum flash evaporation cooling: Modeling. International Journal of Heat and Mass Transfer. 2015;**84**:149-157

[3] Nelson JS, Milner TE, Anvari B, Tanenbaum BS, Kimel S, Svaasand LO, Jacques SL. Dynamic epidermal cooling during pulsed laser treatment of port-wine stain: A new methodology with preliminary clinical evaluation. Archives of Dermatology. 1995; **131**(6):695-700

[4] Zhou Z-F, Chen B, Wang R, Bai F-L, Wang G-X. Coupling effect of hypobaric pressure and spray distance on heat transfer dynamics of R134a pulsed flashing spray cooling. Experimental Thermal and Fluid Science. 2016;**70**:96-104

[5] Li D, He YL, Wang GX. Thermal Modelling for Laser Treatment of Port Wine Stains. Developments in Heat Transfer. InTech; 2011

[6] Aguilar G, Choi B, Broekgaarden M, Yang O, Yang B, Ghasri P, Chen JK, Bezemer R, Nelson JS, van Drooge AM, Wolkerstorfer A, Kelly KM, Heger M. An overview of three promising mechanical, optical, and biochemical engineering approaches to improve selective photothermolysis of refractory port wine stains. Annals of Biomedical Engineering. 2012;**40**:486-506

[7] Chen JK, Ghasri P, Aguilar G, van Drooge AM, Wolkerstorfer A, Kelly KM, Heger M. An overview of clinical and experimental treatment modalities for port wine stains. Journal of the American Academy of Dermatology. 2012;**67**(2):289-304

[8] Anderson R, Parrish J. Selective photothermolysis: Precise microsurgery by selective absorption of pulsed radiation. Science. 1983;**220**:524-527

[9] Arndt KA. Lasers in Cutaneous and Aesthetic Surgery. Lippincott-Raven Publishers; 1997

[10] Aguilar G, Majaron B, Verkruysse W, Zhou Y, Nelson JS, Lavernia EJ. Theoretical and experimental analysis of droplet diameter, temperature, and evaporation rate evolution in cryogenic sprays. International Journal of Heat and Mass Transfer. 2001;**44**:3201-3211

[11] Wu W-T, Yang F, Antaki JF, Aubry N, Massoudi M. Study of blood flow in several benchmark micro-channels using a two-fluid approach. International Journal of Engineering Science. 2015;**95**:49-59

[12] Wu W-T, Aubry N, Massoudi M, Kim J, Antaki JF. A numerical study of blood flow using mixture theory. International Journal of Engineering Science. 2014;**76**:56-72

[13] Sher E, Bar-Kohany T, Rashkovan A. Flash-boiling atomization. Progress in Energy and Combustion Science. 2008;**34**:417-439

[14] Suh PB, Yong LS. An experimental investigation of the flash atomization mechanism. Atomization and Sprays. 1994;**4**:159-179

[15] Peter EM, Takimoto A, Hayashi Y. Flashing and shattering phenomena of superheated liquid jets. JSME International Journal Series B: Fluids and Thermal Engineering. 1994;**37**:313-321

[16] Oza RD. On the mechanism of flashing injection of initially subcooled fuels. Journal of Fluids Engineering-Transactions of the ASME. 1984;**106**:5-9

[17] Wu S, Xu M, Hung DLS, Pan H. In-nozzle flow investigation of flash boiling fuel sprays. Applied Thermal Engineering. 2016;**117**:644-651

[18] Wang X-S, Chen B, Wang R, Xin H, Zhou Z-F. Experimental study on the relation between internal flow and flashing spray characteristics of R134a using straight tube nozzles. International Journal of Heat and Mass Transfer. 2017;**115**:524-536

[19] Pikkula BM, Torres JH, Tunnell JW, Anvari B. Cryogen spray cooling: Effects of droplet size and spray density on heat removal. Lasers in Surgery and Medicine. 2001;**28**:103-112

[20] Pikkula BM, Tunnell JW, Chang DW, Anvari B. Effects of droplet velocity, diameter, and film height on heat removal during cryogen spray cooling. Annals of Biomedical Engineering. 2004;**32**:1131-1140

[21] Karapetian E, Aguilar G, Kimel S, Lavernia EJ, Nelson JS. Effects of mass flow rate and droplet velocity on surface heat flux during cryogen spray cooling. Physics in Medicine and Biology. 2003;**48**:N1-N6

[22] Aguilar G, Majaron B, Pope K, Svaasand LO, Lavernia EJ, Nelson JS. Influence of nozzle-to-skin distance in cryogen spray cooling for dermatologic laser surgery. Lasers in Surgery and Medicine. 2001;**28**:113-120

[23] Aguilar G, Majaron B, Karapetian E, Lavernia EJ, Nelson JS. Experimental study of cryogen spray properties for application in dermatologic laser surgery. IEEE Transactions on Biomedical Engineering. 2003;**50**:863-869

[24] Vu H, García-Valladares O, Aguilar G. Vapor/liquid phase interaction in flare flashing sprays used in dermatologic cooling. International Journal of Heat and Mass Transfer. 2008;**51**:5721-5731

[25] Zhifu Z, Weitao W, Bin C, Guoxiang W, Liejin G. An experimental study on the spray and thermal characteristics of R134a two-phase flashing spray. International Journal of Heat and Mass Transfer. 2012;**55**:4460-4468

[26] Zhou Z-F, Chen B, Wang R, Wang G-X. Comparative investigation on the spray characteristics and heat transfer dynamics of pulsed spray cooling with volatile cryogens. Experimental Thermal and Fluid Science. 2017;**82**:189-197

[27] Yildiz D, Rambaud P, Beeck JV, Buchlin J-M. Evolution of the spray characteristics in superheated liquid jet atomization in function of initial flow conditions. In: 10th International Conference on Liquid Atomization and Spray Systems; Kyoto, Japan. 2006

[28] Yildiz D, van Beeck JPAJ, Riethmuller ML. Feasibility exploration of laser-based techniques for characterization of a flashing jet. Particle & Particle Systems Characterization. 2004;**21**:390-402

[29] Yildiz D, Rambaud P, Beeck JV. Break-up, droplet size and velocity characterizations of a two-phase flashing R134a Jet. In: 5th International Conference on Multiphase Flow; Yokohama, Japan. 2004

[30] Yildiz D, Van Beeck J, Riethmuller M. Global rainbow thermometry applied to a flashing two-phase R134-A jet. In: Proceedings of the Eleventh International Symposium on Application of Laser Techniques to Fluid Mechanics. Citeseer; 2002

[31] Yildiz D, Rambaud P, Van Beeck J, Buchlin J. Thermal characterization of a R134A two-phase flashing jet. In: 9th International Conference on Liquid Atomization and Spray Systems ICLASS. 2003

[32] Polanco G, Holdo AE, Munday G. General review of flashing jet studies. Journal of Hazardous Materials. 2010;**173**:2-18

[33] Zhou Z-F, Lu G-Y, Chen B. Numerical study on the spray and thermal characteristics of R404A flashing spray using OpenFOAM. International Journal of Heat and Mass Transfer. 2018;**117**:1312-1321

[34] Wang R, Chen B, Wang X-S. Numerical simulation of cryogen spray cooling by a three-dimensional hybrid vortex method. Applied Thermal Engineering. 2017;**119**:319-330

[35] Zhou Z-F, Xu T-Y, Chen B. Algorithms for the estimation of transient surface heat flux during ultra-fast surface cooling. International Journal of Heat and Mass Transfer. 2016;**100**:1-10

[36] Aguilar G, Wang GX, Nelson JS. Effect of spurt duration on the heat transfer dynamics during cryogen spray cooling. Physics in Medicine and Biology. 2003;**48**:2169-2181

[37] Pikkula BM, Tunnell JW, Anvari B. Methodology for characterizing heat removal mechanism in human skin during cryogen spray cooling. Annals of Biomedical Engineering. 2003;**31**:493-504

[38] Torres JH, Tunnell JW, Pikkula BM, Anvari B. An analysis of heat removal during cryogen spray cooling and effects of simultaneous airflow application. Lasers in Surgery and Medicine. 2001;**28**:477-486

[39] Tunnell JW, Torres JH, Anvari B. Methodology for estimation of time-dependent surface heat flux due to cryogen spray cooling. Annals of Biomedical Engineering. 2002;**30**:19-33

[40] Jia W, Aguilar G, Wang GX, Nelson JS. Heat-transfer dynamics during cryogen spray cooling of substrate at different initial temperatures. Physics in Medicine and Biology. 2004;**49**:5295-5308

[41] Aguilar G, Vu H, Nelson JS. Influence of angle between the nozzle and skin surface on the heat flux and overall heat extraction during cryogen spray cooling. Physics in Medicine and Biology. 2004;**49**:N147-N153

[42] Franco W, Liu J, Romero-Mendez R, Jia W, Nelson JS, Aguilar G. Extent of lateral epidermal protection afforded by a cryogen spray against laser irradiation. Lasers in Surgery and Medicine. 2007;**39**:414-421

[43] Franco W, Liu J, Wang GX, Nelson JS, Aguilar G. Radial and temporal variations in surface heat transfer during cryogen spray cooling. Physics in Medicine and Biology. 2005;**50**:387-397

[44] Wang R, Zhou Z, Chen B, Bai F, Wang G. Surface heat transfer characteristics of R404A pulsed spray cooling with an expansion-chambered nozzle for laser dermatology. International Journal of Refrigeration. 2015;**60**:206-216

[45] Zhou Z, Chen B, Wang Y, Guo L, Wang G. An experimental study on pulsed spray cooling with refrigerant R-404a in laser surgery. Applied Thermal Engineering. 2012;**39**:29-36

[46] Tian J-m, Chen B, Zhou Z-f. Methodology of surface heat flux estimation for 2D multilayer mediums. International Journal of Heat and Mass Transfer. 2017:675-687

[47] Aguilar G, Wang GX, Nelson JS. Dynamic behavior of cryogen spray cooling: Effects of spurt duration and spray distance. Lasers in Surgery and Medicine. 2003;**32**:152-159

[48] Majaron B, Kimel S, Verkruysse W, Aguilar G, Pope K, Svaasand LO, Lavernia EJ, Nelson JS. Cryogen spray cooling in laser dermatology: Effects of ambient humidity and frost formation. Lasers in Surgery and Medicine. 2001;**28**:469-476

[49] Ramirez-San-Juan JC, Choi B, Franco W, Nelson JS, Aguilar G. Effect of ambient humidity on light transmittance through skin phantoms during cryogen spray cooling. Physics in Medicine and Biology. 2006;**51**:113-120

[50] Vu H, Aguilar G, Stuart Nelson J. Passive mass deposition control of cryogen sprays through the use of wire meshes. Lasers in Surgery and Medicine. 2004;**34**:329-334

[51] Majaron B, Svaasand LO, Aguilar G, Nelson JS. Intermittent cryogen spray cooling for optimal heat extraction during dermatologic laser treatment. Physics in Medicine and Biology. 2002;**47**:3275-3288

[52] Basinger B, Aguilar G, Nelson JS. Effect of skin indentation on heat transfer during cryogen spray cooling. Lasers in Surgery and Medicine. 2004;**34**:155-163

[53] Tian JM, Chen B, Li D, Zhou ZF. Transient spray cooling: Similarity of dynamic heat flux for different cryogens, nozzles and substrates. International Journal of Heat and Mass Transfer. 2017;**108**:561-571

[54] Tuqan AT, Kelly KM, Aguilar G, Ramirez-San-Juan JC, Sun CH, Cassarino D, Derienzo D, Barr RJ, Nelson JS. Evaluation of single versus multiple cryogen spray cooling spurts on in vitro model human skin. Lasers in Medical Science. 2005;**20**:80-86

[55] Kao B, Kelly KM, Aguilar G, Hosaka Y, Barr RJ, Nelson JS. Evaluation of cryogen spray cooling exposure on in vitro model human skin. Lasers in Surgery and Medicine. 2004;**34**:146-154

[56] Li D, Chen B, Wu WJ, Wang GX, He YL. Multi-scale modeling of tissue freezing during cryogen spray cooling with R134a, R407c and R404a. Applied Thermal Engineering. 2014;**73**:1489-1500

[57] Li D, Chen B, Wu WJ, He YL, Xing LZ, Wang GX. Numerical analysis of cold injury of skin in cryogen spray cooling for laser dermatologic surgery. In: ASME International Mechanical Engineering Congress & Exposition Proceedings. 2014. pp. 673-681

[58] Dai T, Yaseen MA, Diagaradjane P, Chang DW, Anvari B. Comparative study of cryogen spray cooling with R-134a and R-404a: Implications for laser treatment of dark human skin. Journal of Biomedical Optics. 2006;**11**:041116

[59] Zhou ZF, Chen B, Wang GX. Experimental and theoretical study on the thermal characteristics of flashing spray cooling using R404a. In: The International Heat Transfer Conference. 2014

[60] Wang R, Chen B, Wang J, Tian J, Zhou Z, Ying Z. Experimental research of R1234yf transient spray cooling and influence of cryogen superheat degree. CIESC Journal. 2018;**69**:595-601

[61] Aguilar G, Franco W, Liu J, Svaasand LO, Nelson JS. Effects of hypobaric pressure on human skin: Implications for cryogen spray cooling (part II). Lasers in Surgery and Medicine. 2005;**36**:130-135

[62] Aguilar G, Svaasand LO, Nelson JS. Effects of hypobaric pressure on human skin: Feasibility study for port wine stain laser therapy (part I). Lasers in Surgery and Medicine. 2005;**36**:124-129

[63] Childers MA, Franco W, Nelson JS, Aguilar G. Laser surgery of port wine stains using local vacuum pressure: Changes in skin morphology and optical properties (part I). Lasers in Surgery and Medicine. 2007;**39**:108-117

[64] Zhou Z-F, Wang R, Chen B, Yang T, Wang G-X. Heat transfer characteristics during pulsed spray cooling with R404A at different spray distances and back pressures. Applied Thermal Engineering. 2016;**102**:813-821

[65] Wang XS, Chen B, Zhou ZF. Atomization and surface heat transfer characteristics of cryogen spray cooling with expansion-chambered nozzles. International Journal of Heat and Mass Transfer. 2018;**121**:15-27

Chapter 8

Water as a Refrigerant in Centrifugal Compressor Cooling Systems for Industrial Applications

Florian Hanslik and Juergen Suess

Additional information is available at the end of the chapter

http://dx.doi.org/10.5772/intechopen.79614

Abstract

As a consequence of the F-gas regulation R404A is no longer a viable option for commercial refrigeration applications. Therefore, this paper focuses on natural refrigerants. There are a few alternatives like carbon dioxide, which has an efficiency loss with increasing environment temperatures. A promising option is the combination of a carbon dioxide process with a chiller using water as refrigerant. Two types of interconnection seem to make energy sense. On the one hand, the interconnection as a cascade, whereby the complete heat of condensation is given off to the water chiller, on the other hand the subcooling of transcritical CO_2 after the gas cooler. Both types of interconnection result in optimized operating parameters for the CO_2 process. These are examined more closely, and finally, the annual COP values are compared with the standard systems.

Keywords: refrigeration, water, carbon dioxide cascade, subcooling, vapor compression, chiller, energy efficiency

1. Introduction

The F-gas regulation, which came into force on January 1, 2015, envisages an EU-wide phase-down of the CO_2 equivalent of hydrofluorocarbons (HFCs) by 79% by 2030 compared to a reference value based on the annual average of the quantities of hydrofluorocarbons, a producer or importer reported to have placed in the market between 2009 and 2012. From January 1, 2020, the next step will be a ban of placing refrigeration appliances for commercial use with HFCs with GWP > 2500 [1]. As a result, the refrigerant blend R404A may no longer be used in new systems from this point in time.

A current alternative is HFC or HFO replacement, but in most cases, they are only seen as a temporary solution. A long-term alternative is natural refrigerant. Water stands out as a natural refrigerant because it has no GWP or ODP and is neither flammable nor toxic. When used in a refrigeration system, it occurs in the liquid and gaseous states. The thermodynamic process takes place in a rough vacuum due to the vapor pressure curve of water but then corresponds to the cycle of conventional refrigeration systems.

In this chapter, the properties of water as a refrigerant are considered in more detail, an overview of previously implemented systems is given, and the mode of operation and special features of such a system are explained. Using examples, the promising field of applications gets defined, and the efficiency of the systems is shown.

2. Water as refrigerant

2.1. The thermodynamic properties of water

Water (R718) as a refrigerant is one of the oldest fluids being used for refrigeration applications down to about the freezing point. When water is coupled with protective solutions to prevent freezing (i.e., propylene or ethylene glycol), it can also be used well below water's normal freezing point in applications such as ice slurries. Water is easily available and has excellent thermodynamics and chemical properties. It was found by reference [2] that for evaporator temperatures above 20°C R718 gives the highest COP of all refrigerants assuming exactly the same cycle parameters [3].

Figure 1 shows the vapor pressure curves of various refrigerants. In the case of R718, it is noticeable that the pressures in the temperature range for a typical cooling application in server rooms are between 10 and 100 mbar.

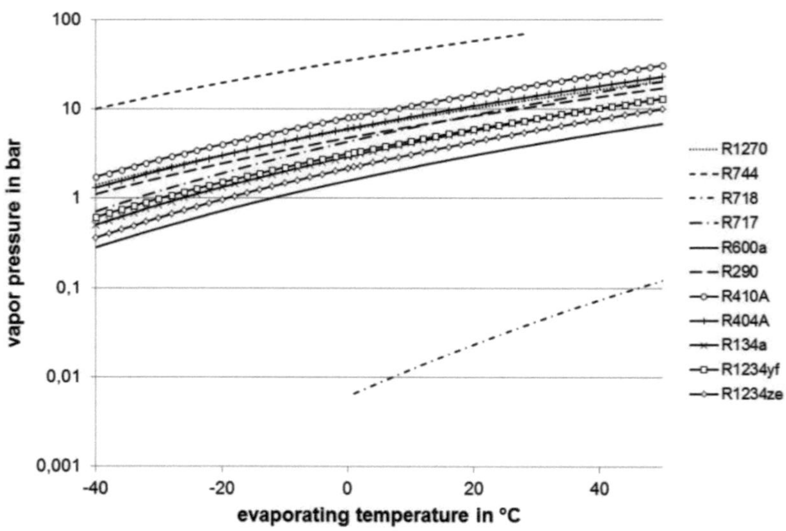

Figure 1. Vapor pressure curves of selected refrigerants.

In addition to the necessary operation in a rough vacuum, **Figures 2** and **3** show the low volumetric cooling capacity and the high required pressure ratio of water as a refrigerant compared to conventional refrigerants.

Both properties require an optimal compressor concept as well as optimum flow control of the water vapor, as this is the only way to minimize the possible losses and to use the advantages of water as a refrigerant efficiently.

The curves of the respective refrigerants in **Figures 1–3** were calculated using REFPROP [4].

2.2. Compressor cooling systems for the refrigerant water

Up to now, quite a few centrifugal chillers were built with R718 as working fluid. In **Table 1** the most relevant projects are listed. The cooling capacity of the majority of the machines is typically above 500 kW, and all machines use centrifugal compressors with the exception of the chiller described by reference [5]. Here, a positive displacement compressor of the Roots blower type is applied. The centrifugal compressors are either of the axial type with up to seven stages or use radial impeller designs with typically two stages. In most machines inter-cooling between the stages is realized, as water vapor has a rather high discharge temperature due to its large polytropic index of approximately 1.33 in the operational range of the compressor [3].

While R718 as refrigerant is state of the art in absorption chillers with lithium bromide as absorbent, none of the mechanical water vapor compression machines shown in **Table 1** have reached a commercially viable status so far, and their market penetration is still pending. According to Ophir [6], only IDE Technologies which is an Israelian water desalination company has successfully built and commissioned a significant number of installations in the field, which still cannot be regarded as industrialized mass-produced machines [3].

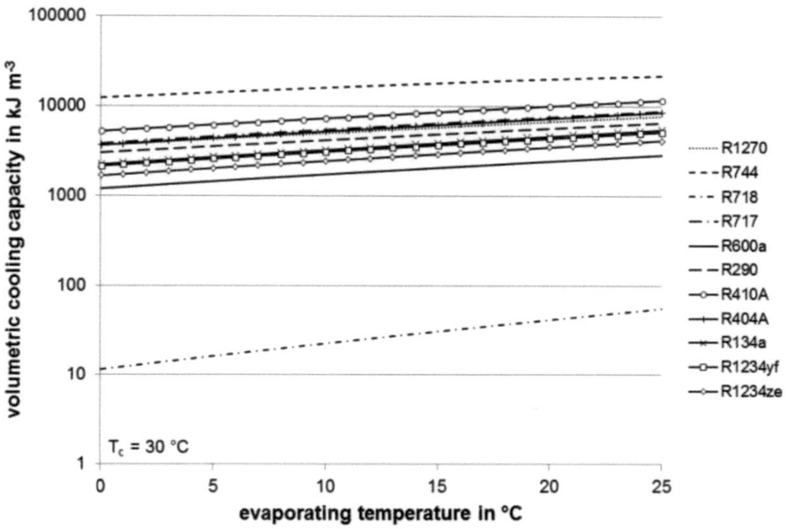

Figure 2. Volumetric cooling capacities of selected refrigerants.

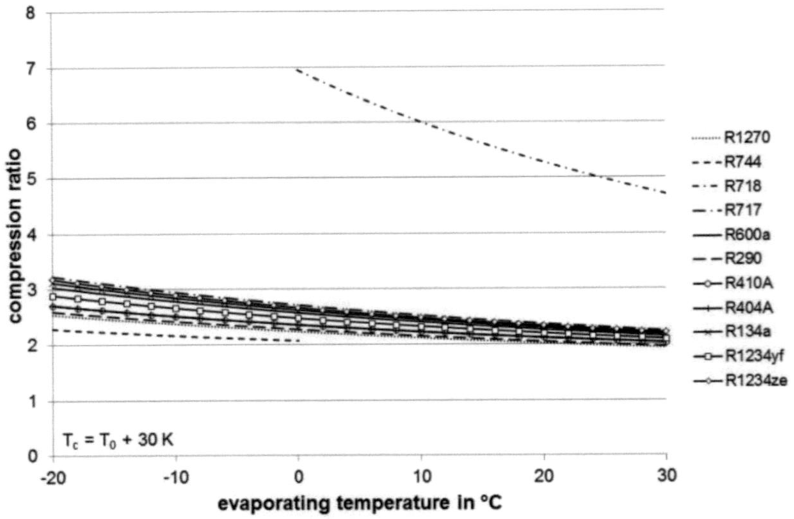

Figure 3. Required compression ratios of selected refrigerants.

Year	Location	Project group	Application	Cooling capacity	Literature source
1990	Denmark	IDE/Sabroe	Sea water heat pump	1.3 MW	[6]
1995	Denmark	IDE/Sabroe/TI	Plastic molding cooling	2 MW	[6]
1995	RSA	IDE/Integral	Gold mine cooling	6 × 3 MW	[6]
1999	Germany	ILK/GHS Essen	Chilled water	1 MW	[7]
1999	Germany	ILK/VW	Chilled water	1 MW	[8]
2000	Germany	ILK/Daimler	Chilled water	1 MW	[7]
2002	Japan	IDE/Sanken	Vacuum ice	350 kW	[6]
2007	Japan	IDE/Nissan	Vacuum ice	1.5 MW	[6]
2008	RSA	IDE	Gold mine cooling	3 × 3 MW	[6]
2008	Switzerland	IDE	Artificial snow	1.4 MW	[6]
2008	Austria	IDE	Artificial snow	1.4 MW	[6]
2011	USA	MSU	Chilled water	—	[9]
2012	Denmark	Rotrex/Topsøe	Heat pump	350 kW	[10]
2013	Japan	Kawasaki HI	Chilled water	350 kW	[11]
2012	Germany	ILK/Cofely Ref.	Chilled water	800 kW	[12]
2014	Japan	Sasakura	Chilled water	17.5 kW	[5]
2015	Denmark	TI/JCI	Chilled water	800 kW	[13]
2015	Germany	ILK	Vacuum ice	50 kW	[12]
2015	Germany	efficient energy	Chilled water	35 kW	[14]

Table 1. Centrifugal chiller projects using R718 as refrigerant [3].

3. Centrifugal compressor cooling system

The functional and working process of a compact compression refrigeration system with the refrigerant water is explained below.

The chiller fulfills two tasks: on the one hand, the cooling of the chilled water to the desired cold water outlet temperature and, on the other hand, the removal of the heat extracted from the water from a heat source. Of importance are the two identical refrigeration modules, one of which is shown in **Figure 4**. Each refrigeration module comprises the complete thermodynamic cycle. The evaporator and the condenser are nested to achieve a compact design. The chilled water enters the evaporator via the middle connection of the refrigeration module at a flow rate of approx. 2 liters per second. Here, there is a pressure that corresponds to the saturated pressure of the desired water outlet temperature of the evaporator. It is evaporated so much water until the superheated water has cooled to saturation temperature. The chilled cold water is directed downward out of the module. The maximum resulting vapor mass corresponds to about 1% of the circulated mass flow of water. The vapor enters the centrifugal compressor and is compressed to a higher pressure and temperature. The maximum achievable pressure ratio is currently in the range of 3.5 due to the rotational speed limitation of the compressors. The compression starts at approx. 40,000 revolutions per minute with a pressure ratio of 1 and can be continuously increased from there up to the maximum pressure ratio. The compressed, superheated steam flows into the condenser where it meets the cooling water. The cooling water is also fed to the cooling module from below. Due to the special design of the condenser, it is achieved that the water vapor can deliver its complete energy

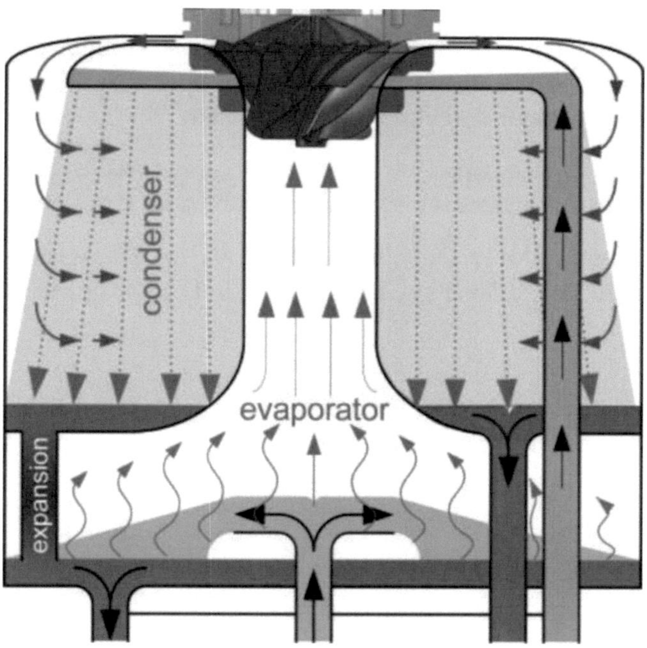

Figure 4. Schematic drawing of the cooling module.

to the circulating cooling water and gets completely condensed. The heated cooling water is discharged down from the module. To close the thermodynamic cycle, the evaporated amount of water from the condenser is returned via the self-regulating expansion device in the evaporator.

Depending on the heat sink temperature, this process can be followed by the second cooling module as a cascade.

4. Combination with water-loop cooled self-contained units

This subsection investigates the combination of a CO_2 process of a self-contained unit with water-loop cooling and a compression refrigeration unit with the refrigerant water in a cascade arrangement.

4.1. Basics

4.1.1. R744 cycle

The basic structure of a single-stage R744 refrigeration cycle is shown in **Figure 5**. Depending on the heat sink temperature, a distinction is made between a subcritical system and a transcritical system. The mode of operation determines the selection of components to a considerable extent.

4.1.2. R718 cycle

The properties of R718 shown in Chapter 2.1 set the following requirements for the use of water as a refrigerant. The low vapor pressures require operation in vacuum, and the large pressure ratio requires a two-stage compression in cascade.

An overview of the machines implemented and the technical challenges of using water as a refrigerant in a centrifugal compressor cooling system have been described by reference [3].

Figure 6 shows a one-step process for both systems (a) and (b). This is for simplification of the illustration only. In the later calculation, however, the real two-stage operation is considered.

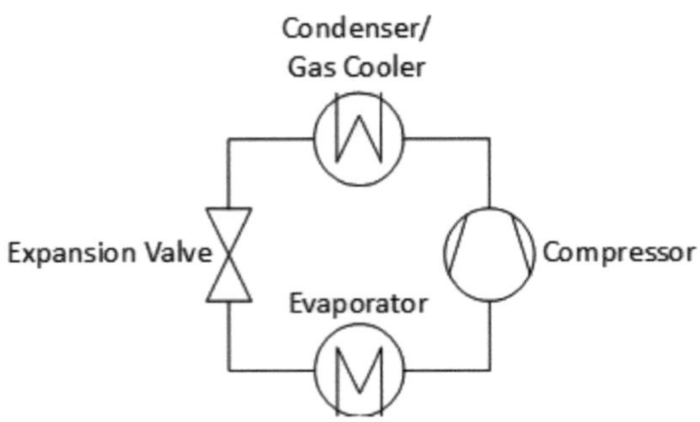

Figure 5. Schematic of R744 refrigeration system.

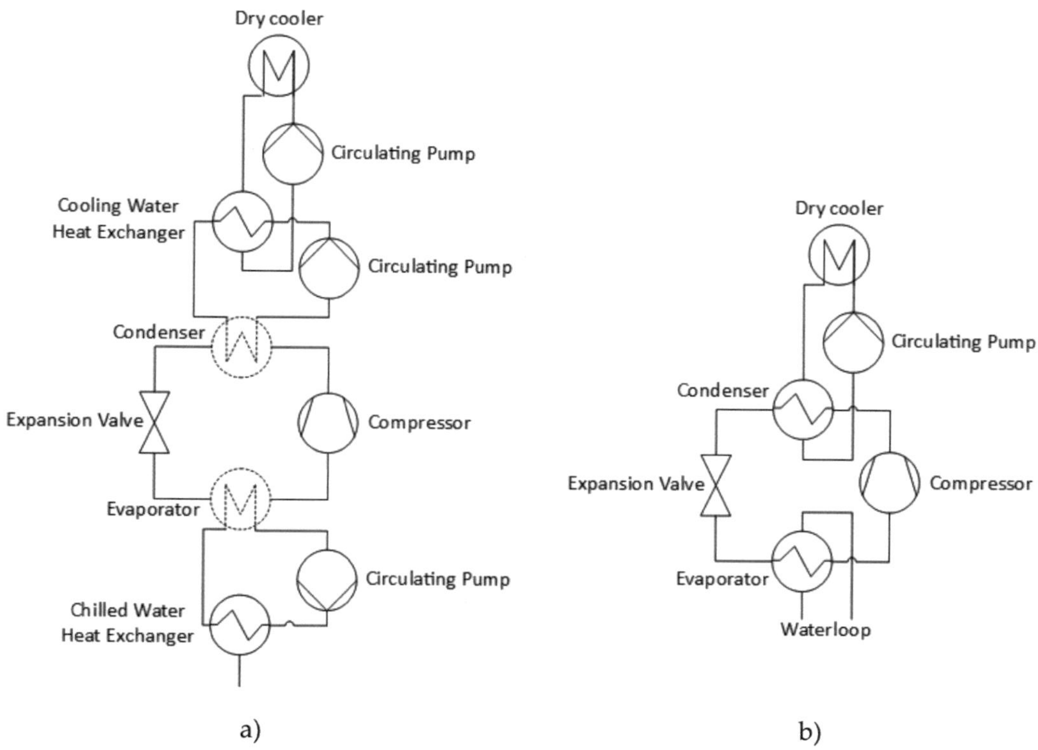

Figure 6. (a) Schematic of a R718 refrigeration system with external heat exchangers (without i.c.). (b) Schematic of a R718 refrigeration system with integrated heat exchangers (with i.c.).

System (a) is a two-stage design with two identical refrigeration modules, with the first-stage condenser connected to the second-stage evaporator by a water interface. The required water circulation is generated by a circulating pump. Both condenser and evaporator are designed as open systems and, hence, the adapted, dashed symbol in the flow diagram. The heat transfer on the heat source as well as the heat sink side of the refrigeration system is done by braze plate heat exchangers. Again, a separate circulation pump is used for each circuit. Thus, a total of three pumps are required to operate the system in its present design stage.

System (b) differs from system (a) in the following points. The intermediate circle is not realized in this structure by a water interface but by a vapor interface with intermediate cooling (i.c.). Furthermore, the heat exchangers are integrated in the evaporator and condenser. Therefore, the three required internal circulation pumps in the system are obsolete.

4.1.3. Possible designs and combination of cascaded refrigeration processes

Table 2 lists the three system combinations under consideration. System 1 is a coupled system of a self-contained unit with a CO_2 compression refrigeration system, a water-loop and a dry cooler outside the building. The condenser/gas cooler of the CO_2 system is designed as a braze plate heat exchanger and represents the interface to the water-loop. The condensing heat is transported by means of the water-loop to the dry cooler, and this dissipates

System 1	CO_2 + water loop
System 2	CO_2 + water loop + R718 chiller without i.c. + external heat exchangers
System 3	CO_2 + water loop + R718 chiller with i.c. + integrated heat exchangers

Table 2. The different systems.

the waste heat to be ambient. In this system, the CO_2 process is subcritical or transcritical depending on the ambient temperature. As a result, all components must be designed for transcritical operation and the compressor for maximum load. These are then oversized for most of the year.

Systems 2 and 3 each consist of a self-contained unit with a CO_2 compression refrigeration system, a water-loop and an additional centrifugal compressor cooling system with water as refrigerant. This ensures that the CO_2 process can always be subcritical, as the water-loop always generates a constant water temperature for condensing CO_2 which is below the critical temperature of CO_2. The differences between Systems 2 and 3 are in the type of the R718 system used. System 2 is shown in **Figure 6(a)** and System 3 in **Figure 6(b)**.

The reference for the comparison is System 1.

4.2. Methods

For the qualitative assessment, the three systems were calculated with the following assumptions (**Table 3**). The physical properties for the cycle process calculation of the R718 process were generated using REFPROP [4].

R744		
Superheating	10	K
Cooling capacity (all systems)	22	kW
Evaporation temperature T_0	−5/−15	°C
$T_{GC/C}$−T_{amb}	4	K
R718		
Chilled water temperature	13.5/24.5	°C
Compressor efficiency	0.6	—
Electrical efficiency	0.8	—
$T_{c.w.}$−T_{amb}	4	K
Combination		
Maximum ambient temperature	35	°C
$T_{GC/C}$−T_{WL}	1.5	K

Table 3. Operating conditions.

The electrical power consumption of the R744 compressor was determined for all three systems by means of the manufacturer's design software [15] for the respective operating points. For all systems a constant power consumption of 400 W was assumed for each water-loop circulation pump. System 2 also took into account the electrical power consumption of the three internal circulation pumps with a constant power consumption of 500 W and the electrical power consumption of the two R718 centrifugal compressors. With System 3, the three pumps are eliminated, and thus only the two centrifugal compressors have been considered.

The electrical power consumption of the fans on the dry cooler or condenser/gas cooler has not been taken into account for the qualitative comparison of calculations in any system.

The exact calculations of the individual systems can be found in Formulas 1–3.

$$COP_1 = \frac{Q_0}{P_{R744} + P_{WL}} \tag{1}$$

$$COP_2 = \frac{Q_0}{P_{R744} + P_{WL} + P_{Pumps} + P_{R718}} \tag{2}$$

$$COP_3 = \frac{Q_0}{P_{R744} + P_{WL} + P_{R718}} \tag{3}$$

4.3. Results and comparison of the systems

As described in Section 2.4, System 1 is designed according to the maximum ambient temperature. Systems 2 and 3 are calculated with two condensing temperatures, 15 and 26°C. The required piston displacements of the carbon dioxide compressors for the indicated cooling capacity of 22 kW at −5 and −15°C evaporation temperature can be taken from **Table 4** for all three systems. The compressors were designed with frequency-controlled motors.

The required piston displacements are lower by a factor of two for Systems 2 and 3 than for System 1.

Figures 7 and **8** show the COP characteristics of the three systems above the ambient temperature. System 1 is used as the reference for the comparison of the systems, as described in Section 2.4. From both figures, it gets obvious that the switching point for the optimum energetic mode of operation depends on two factors:

- The evaporation temperature of the CO_2 process
- The condensation temperature of the CO_2 process

Systems	**1**	**2/3**	
	T_C = 40.5°C	T_c = 26°C	T_c = 15°C
T_0 = −5°C	9.6 $m^3 h^{-1}$	6.5 $m^3 h^{-1}$	4.8 $m^3 h^{-1}$
T_0 = −15°C	12 $m^3 h^{-1}$	9.6 $m^3 h^{-1}$	6.5 $m^3 h^{-1}$

Table 4. The required piston displacements of the carbon dioxide compressors depending on the system.

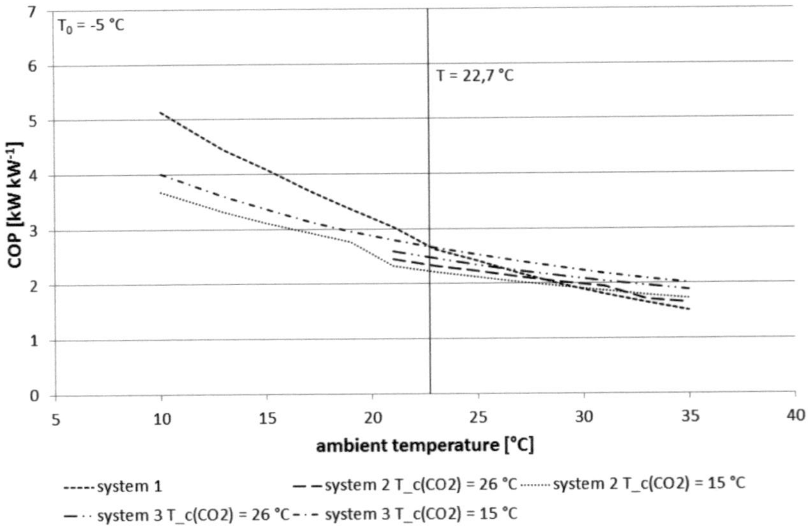

Figure 7. Comparison of the COP values at $T_0 = -5°C$.

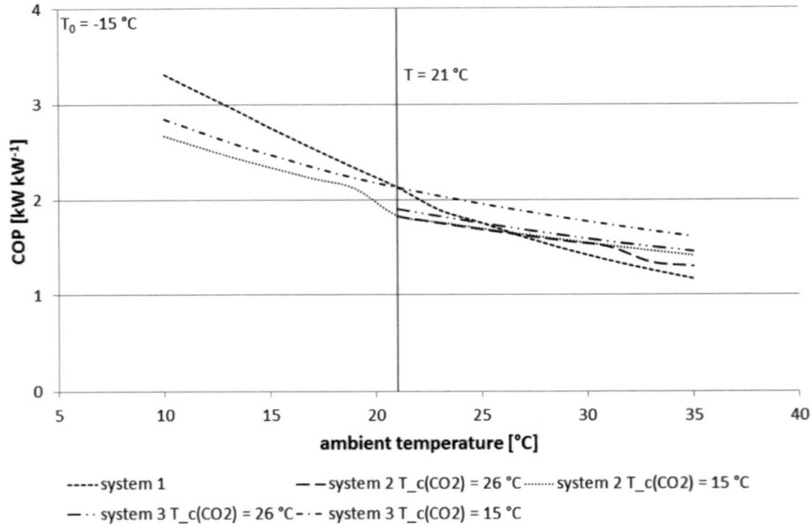

Figure 8. Comparison of the COP values at $T_0 = -15°C$.

For both factors the statement applies, the lower, the better.

Thus, the switching point for $T_0 = -5°C$ and $T_C = 15°C$ at $T_{amb} = 22.7°C$ and for $T_0 = -15°C$ and $T_C = 15°C$ at $T_{amb} = 21°C$ compared with System 1. In both cases, System 3 has a significantly better energetic performance than System 2. This can be explained by the fact that in System 3 the circulating pumps of the R718 process can be dispensed with, and the losses in the water cycle between the two refrigeration modules are eliminated by using the intermediate cooling.

5. Mechanical subcooling of transcritical carbon dioxide

5.1. The investigated system

Figure 9 shows the schematic structure of the investigated system. It is a single-stage CO_2 cycle, the "refrigeration cycle," and a subcooling circuit "mechanical subcooling" downstream of the gas cooler. The aim is to further cool the transcritical CO_2 leaving the gas cooler by means of an additional compression refrigeration system. The refrigerant in this subcooling circuit is R718. The interface between the two circuits is a finned tube heat exchanger, which is traversed by CO_2 inside and is surrounded by circulating water in a vacuum atmosphere. The energy required for the evaporation of the water is taken from the CO_2 gas, thereby cooling it. The resulting water vapor is compressed by means of a centrifugal compressor and fed into the condenser. There, the water vapor rejects its energy to another finned tube heat exchanger to an additional cooling circuit and condenses completely. The circuit is then closed by a self-regulating, pressure loss-free expansion device. The expansion device used in combination with the continuously variable centrifugal compressor allows a continuous adjustment of the delivered volume flow, and the ratio between the condensing pressure and suction pressure of the compressor starts as low as at a value of "1."

The additional cooling circuit in the considered system consists of the heat exchanger, a circulation pump, and a dry cooler. As a working medium, a glycol/water brine is usually used. This extra circuit is needed because there are no commercially available air condensers for

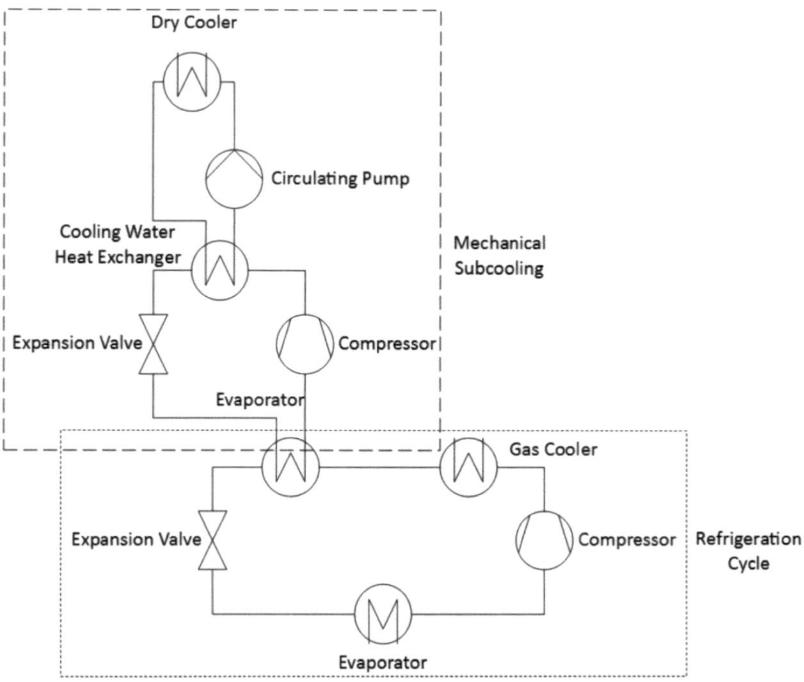

Figure 9. Schematic of the combined subcooling cycle.

R718. The problem is the given density ratio of >10,000, at a temperature of 50°C increasingly with decreasing water or steam temperature. Both systems, gas coolers and dry coolers, transfer their waste heat to the same heat sink, the environment.

5.2. Method

For the evaluation of the system with and without mechanical subcooling, the assumptions given in **Table 5** were used as the basis for the calculations. The physical properties of the refrigerants used for the respective cycles were generated with REFPROP [4].

For the calculation of the individual COP values, the electrical power consumption of the circulation pump in the external cooling water circuit of the R718 chiller as well as the fans of the gas cooler and the dry cooler were neglected. Only the specific capacities were considered. Eq. (4) shows the general calculation of COP, which is also used for the determination of pure transcritical operation. q_0 corresponds to the specific cooling capacity and w_c to the required specific compressor work of the refrigeration cycle.

$$COP = \frac{q_0}{w_c} \tag{4}$$

Eq. (5) shows the calculation of the specific cooling capacity and Eq. (6) the specific subcooling capacity of the CO_2 cycle. Eq. (7) shows the specific cooling capacity of the R718 circuit. h_0 and h_5 correspond to the specific enthalpy at the outlet of the evaporator or after the throttle, h_3 and h_4 to the specific enthalpy at the outlet of the gas cooler or after subcooling in the CO_2 cycle, and h_{1*} and h_{4*} to the specific enthalpy in the evaporator or after relaxing in the R718 circle.

$$q_{0,R744} = h_0 - h_5 \tag{5}$$

$$q_{sub} = h_3 - h_4 \tag{6}$$

$$q_{0,R718} = h_{1*} - h_{4*} \tag{7}$$

The energy balance of the subcooler is shown in Eq. (8), and Eq. (9) shows the relation of the occurring mass flows.

$$\dot{m}_{R744} * q_{sub} = \dot{m}_{R718} * q_{0,R718} \tag{8}$$

$$\dot{m}_{R718} = \frac{\dot{m}_{R744} * q_{sub}}{q_{0,R718}} \tag{9}$$

The specific compressor work of the two single-stage systems is shown in Eq. (10) for the R744 cycle and in Eq. (11) for the R718 process. h_1 and h_{1*} represent the specific enthalpy at the compressor inlet, h_2 and h_{2*} the isentropic specific enthalpy at the compressor outlet. $\eta_{i,R744}$ and $\eta_{i,R718}$ are the isentropic efficiencies of the respective compressors.

R744		
Superheating	10	K
Cooling capacity	150	kW
Evaporating temperature	−5/−15	°C
t_3-t_{env}	5	K
Isentropic compressor efficiency [16]	0.95–0.1*π	—
R718		
t_4-t_{1*}	5	K
compressor efficiency	0.7	—
t_{3*}-t_{env}	4	K
maximum compressor volume flow	1.2	$m^3\ s^{-1}$

Table 5. Operating conditions.

$$w_{c,R744} = \frac{h_{2,s} - h_1}{\eta_{i,R744}} \tag{10}$$

$$w_{c,R718} = \frac{h_{2^*,s} - h_{1^*}}{\eta_{i,R718}} \tag{11}$$

Based on Eq. (4), the COP of the entire system is calculated in subcooling mode according to Eq. (12).

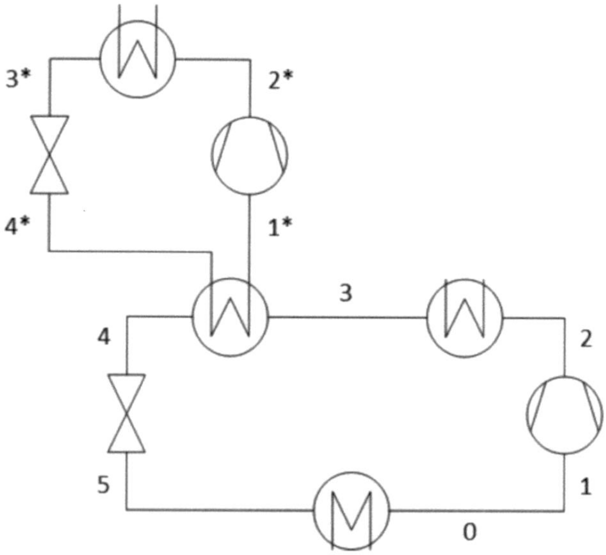

Figure 10. Schematic with state points.

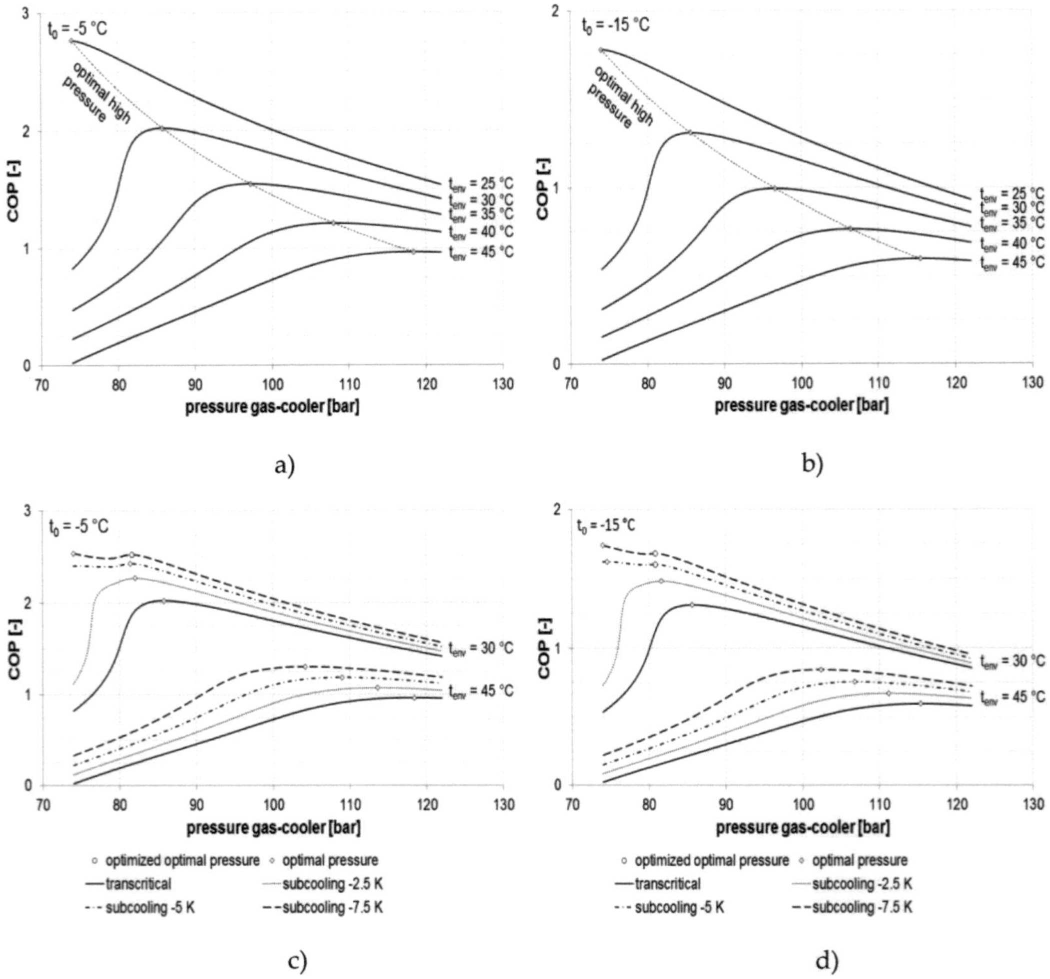

Figure 11. Optimal high pressure for the transcritical CO_2 cycle with (a) t_0 = −5°C and (b) t_0 = −15°C; optimal and optimized pressure with subcooling for (c) t_0 = −5°C and (d) t_0 = −15°C.

$$COP^* = \frac{\dot{m}_{R744} * q_{0,R744}}{\dot{m}_{R744} * w_{c,R744} + \dot{m}_{R718} * w_{c,R718}} = \frac{q_{0,R744}}{w_{c,R744} + \frac{q_{sub}}{q_{0,R718}} * w_{c,R718}} \qquad (12)$$

The individual states of the respective circuits are shown in **Figure 10**.

5.3. Results

5.3.1. Optimum operating conditions

Figure 11 shows the optimal high pressures of the transcritical CO_2 system, (a) for t_0 = −5°C and (b) for t_0 = −15°C, for different ambient temperatures. The respective dashed lines represent the interpolated connections between the individual maximum points. The individual marked values have http://dx.doi.org/10.5772/intechopen.79614

been determined by means of a self-developed simulation. From each of the two diagrams, two curves at the ambient temperatures t_{env} = 30°C and t_{env} = 45°C are considered in more detail, and the optimal pressures for operation with a subcooling of −2.5, −5, and −7.5 K are shown. Diagram (c) shows the values for t_0 = −5°C, and diagram (d) shows the values t_0 = −15°C. In both diagrams, it can be seen that the optimum pressure drops as expected with increasing subcooling value. For (c) and (d), the optimum pressures at t_{env} = 30°C and a subcooling of 7.5 K at 74 bar and at (d) are only slightly higher when cooled by 5 K. This is followed by an increase in efficiency with subsequent increase in pressure, followed by a rise to a turning point. From this, the efficiency of the system continues to fall with further increases in process pressure. These inflection points are referred to in the diagrams as *optimized optimal pressure* and are preferable to the maximum efficiency points, since the efficiency values are only slightly lower and there are advantages for selecting the compressor for the subcooling stage. This can be explained by the p-h diagram shown in **Figure 12**.

The three illustrated cycles each show the transcritical CO_2 cycle for the operating point t_0 = −5°C and t_3 = 35°C. The solid line with the triangle symbols at the respective state points represents the pure transcritical cycle without mechanical subcooling at optimum high pressure. The dotted line with the circle symbols represents the transcritical cycle with a subcooling of 7.5 K at optimum high pressure (74 bars), and the dashed line with the rhombuses represents the transcritical cycle with subcooling at the optimized optimum pressure. The points *3* and *4* for the compared subcooling cycles are each on the same isotherms and represent at *3* the temperature at the gas cooler outlet and at *4* the temperature after the subcooling. Provided that the same cooling capacity is required for both systems, both systems need approximately the same mass flow of CO_2. If one compares the enthalpy difference q_{sub} with optimal and optimized optimal pressure, it clearly shows that the required subcooling performance at optimum pressure is more than a factor of two higher than that of the optimally optimized pressure. This would also result in a larger sizing of the R718 chiller.

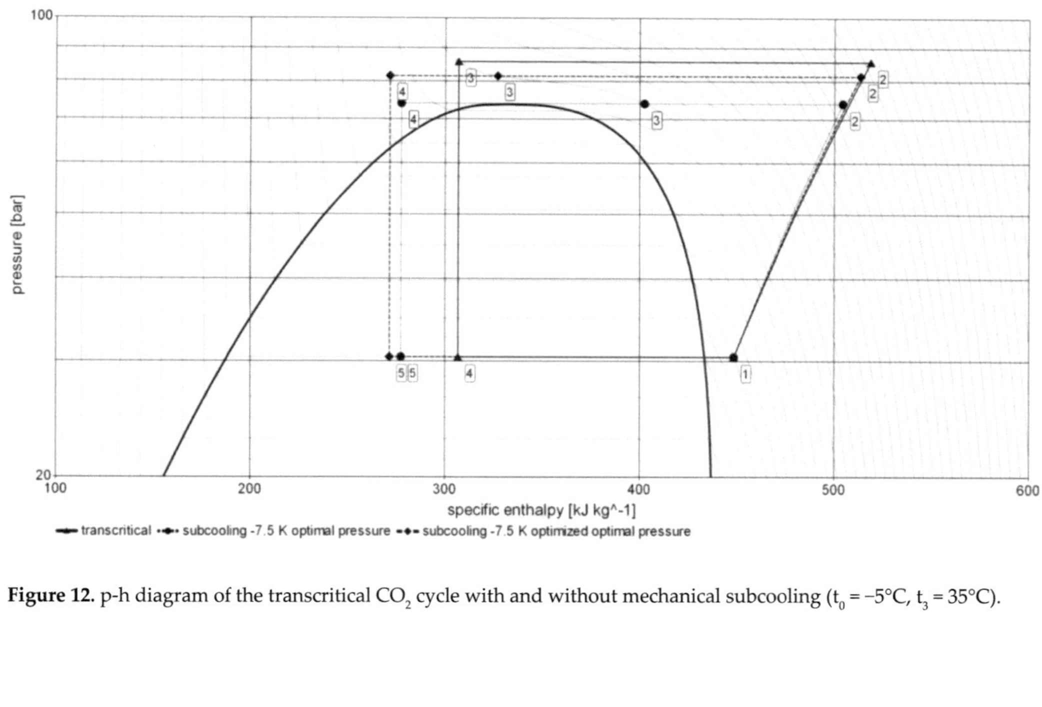

Figure 12. p-h diagram of the transcritical CO_2 cycle with and without mechanical subcooling (t_0 = −5°C, t_3 = 35°C).

5.3.2. *Efficiency increase*

Based on the optimum or optimized optimum operating pressures, **Figure 13** shows the COP curves for pure transcritical operation and for transcritical operation with mechanical subcooling as a function of the environment temperature. Diagram (a) refers to t_0 = −5°C and diagram (b) to t_0 = −15°C. Furthermore, with the respective secondary axis, the efficiency increase between the purely transcritical operation and the operation with a subcooling of 7.5 K is shown. When comparing the two curves, it is noticeable that there is a dependency on the evaporation temperature and the ambient temperature. With decreasing evaporation temperature, as well as with increasing ambient temperature, the percentage increase in efficiency increases. Furthermore, it can be seen that the increase from an ambient temperature of t_{env} = 35°C is significantly lower and seems to approach asymptotically to a maximum limit.

5.3.3. *Required subcooling capacity*

In the following, the required subcooling capacities (SUB) are shown in **Figure 14** with the solid lines, and the maximum possible cooling capacity (Q_0) of the R718 circuit for the three indicated subcooling temperatures is shown by the dashed lines. Diagram (a) refers to t_0 = −5°C and diagram (b) to t_0 = −15°C. The optimized optimum pressure was used as the basis for the calculation.

It is easy to see that for both evaporating temperatures, with a small exception at t_0 = −15°C, with the single-stage R718 system, with the maximum volumetric flow given in **Table 5**, a subcooling of −5 K over the entire temperature range of the environment can be realized. Over a wide range, a subcooling value of >7.5 K is possible with the above setting. Again, for a subcooling of 7.5 K, as in **Figure 13**, a turning point in the curve at t_{env} = 35°C can be seen. In addition, significantly larger subcooling temperatures are possible. Another point is the

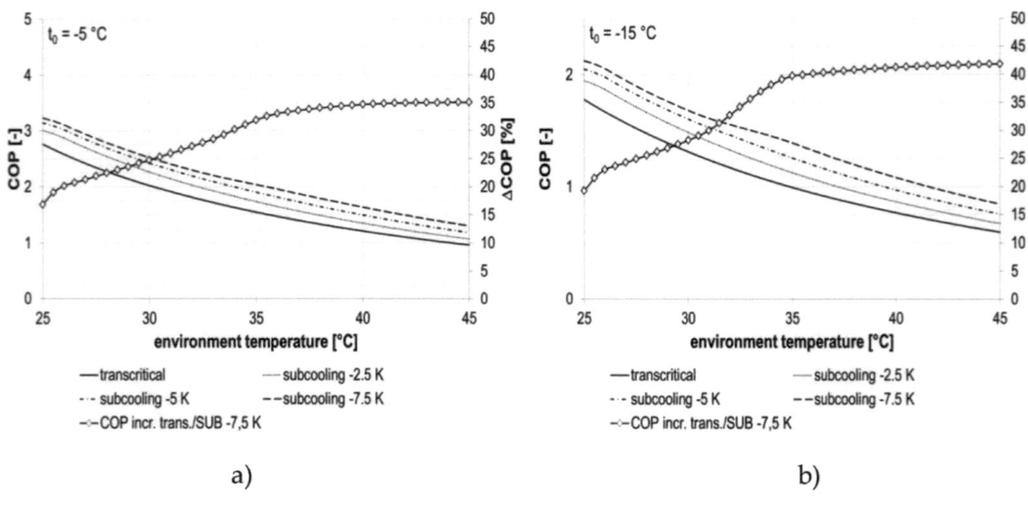

Figure 13. COP of the transcritical CO_2 cycle with and without subcooling: (a) t_0 = −5°C and (b) t_0 = −15°C.

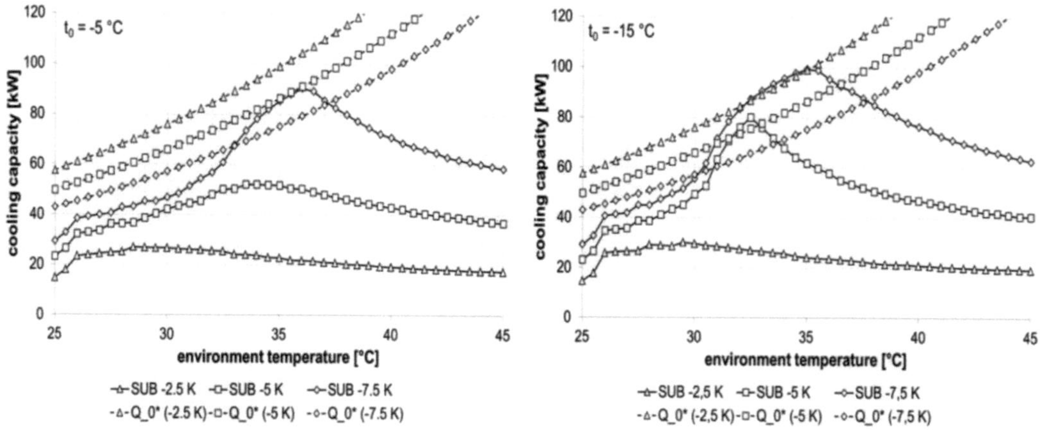

a) b)

Figure 14. Required subcooling capacity and possible cooling capacity: (a) $t_0 = -5°C$ and (b) $t_0 = -15°C$.

increasing possible cooling capacity with higher environment temperatures. This is related to the increase in the density of water vapor as the evaporation temperature increases.

In order to be able to subcool at least 7.5 K over the whole range of the environmental temperatures, there are two options for optimization. The first one could use a R718 compressor with a larger maximum flow rate, and, the second one could increase the process pressure at the outlet of the R744 compressor in order to reduce the required subcooling performance. Both options require further investigation to determine which of the two is more efficient. Furthermore, the combination of the two systems can still be examined to see what absolute subcooling over the entire environment temperature range for the two evaporation temperatures can be achieved.

6. Conclusion

The analysis of the systems has shown that a self-contained unit with water-loop combined with a centrifugal compressor cooling system with water as refrigerant of System 3 at a condensing temperature of $T_C = 15°C$ and an evaporation temperature of $T_0 = -15°C$ of the CO_2 process above an ambient temperature of $T_{amb} = 21°C$ proves to make sense from an energetic point of view. In addition to the energetic consideration, the positive effect of the subcritical design in contrast to System 1 has been noticed. It allows a much simpler design of the CO_2 system and the up to a factor 2 lower required piston displacements of the R744 compressors used.

The simulation of a transcritical CO_2 process with subsequent mechanical subcooling with a refrigeration system with the refrigerant R718 has shown that efficiency increases of more than 35% compared to purely transcritical operation can occur. The main influencing factors regarding the efficiency are on the one hand the evaporation and ambient temperatures and, on the other hand, the process pressure on the pressure side of the compressor. It has been noticed that in the course of the COP curve above the environmental temperature, there are, in addition to the optimal process pressure, also points which have a positive effect on the entire system with a slight loss of efficiency. For both investigated evaporation temperatures in the CO_2 cycle, a subcooling of 5 K is possible with the considered system with a small exception over the entire ambient temperature range. Over much of the environmental temperatures, significantly greater temperature differences are possible. In order to allow a subcooling of 7.5 K over the entire temperature range, further investigations have to be made, which on the one hand consider a larger compressor and on the other hand a further optimized process pressure.

Nomenclature

COP	coefficient of performance (kW kW^{-1})
*COP**	overall coefficient of performance (–)
GWP	global warming potential (–)
h	specific enthalpy (kJ kg^{-1})
$\dot{m}$	mass flow (kg s^{-1})
ODP	ozone depletion potential (–)
P	electrical power consumption (kW)
q_0	specific cooling capacity (kJ kg^{-1})
Q_0	cooling capacity (kW)
q_{sub}	specific subcooling capacity (kJ kg^{-1})
t_0	evaporating temperature (°C)
w_c	specific compression work (kJ kg^{-1})
Greek symbols	
η	compressor efficiency (–)
π	pressure ratio (–)
Subscript	
0…5	condition point of the CO_2 cycle
$1^*…4^*$	condition point of the R718 cycle
env	environment
Pumps	related to circulating pumps

R718	related to the R718 cycle
R744	related to the R744 cycle
WL	related to the water-loop circulating pump

Author details

Florian Hanslik* and Juergen Suess

*Address all correspondence to: florian.hanslik@efficient.energy.de

Efficient Energy GmbH, Feldkirchen, Germany

References

[1] European Commission. Regulation (EU) No 517/2014 of the European Parliament and of the Council of 16 April 2014 on fluorinated greenhouse gases and repealing Regulation (EC) No 842/2006: (OJ L 150 of 20.5.2014, pp. 195-230); 2014

[2] Kilicarslan A, Müller N. A comparative study of water as a refrigerant with some current refrigerants. International Journal of Energy Research. 2005;**29**:947-959. DOI: 10.1002/er.1084

[3] Suess J. A centrifugal compressor cooling system using water as working fluid. In: 12th IIR-Gustav Lorentzen Natural Working Fluids Conference (GL 2016); 21-24 August; Edinburgh, United Kingdom; 2016. pp. 1178-1184. DOI: 10.18462/iir.gl.2016.1204

[4] Lemmon EW, Huber ML, McLinden MO. NIST Reference Fluid Thermodynamic and Transport Properties Database (REFPROP); 2013

[5] Sasakura T. Report 2014 Sasakura Engineering Co., Ltd. 7-32 Takejima 4-chome, Nishi-yodogawa-ku, Osaka 555-0011, Japan

[6] Ophir A. Mechanical heat pumps using water as refrigerant for ice production and air conditioning. In: IDEA 99th Annual Converence & Trade Show, Orlando, Florida; 2008

[7] Albring P. Kältetechnik mit Wasser als Kältemittel, Netzwerk Kältetechnik; 2009

[8] Albring P, Honke M. Ice-making and ice storage, with water as refrigerant. In: 23rd IIR International Congress of Refrigeration; 21-26 August; Prague, Czech Republic; 2011

[9] Li Q, Piechna J, Müller N. Thermodynamic potential of using a counter rotating novel axial impeller to compress water vapor as refrigerant. International Journal of Refrigeration. 2011;**34**:1286-1295. DOI: 10.1016/j.ijrefrig.2011.03.002

[10] Borup J, Jensen J. Udvikling af Rotrex turbokompressor til vanddamp compression, Nr. 344-009, PSO Program 2012, Dansk Energi

[11] Murayama S. Report 2014 Kawasaki Heavy Industries, Ltd. Tokyo Office, 14-5 Kaigan 1-chrome, Minato-ku, Tokyo 105-8315, Japan

[12] Honke M, Safarik M, Herzog R. R718 turbo chillers and vacuum ice generation–Two applications of a new generation of high speed, high capacity R718 centrifugal compressors. In: 24th IIR International Congress of Refrigeration; 16-22 August; Yokohama, Japan; 2015

[13] Madsbøll H. Competitive Chiller Concept with Water as Refrigerant, Danske Køledage; 2013

[14] Süß J. Eine kompakte Kälteanlage mit Wasser als Kältemittel. In: Deutsche Kälte- und Klimatagung; 19-21 November 2014; Düsseldorf; 2014

[15] Bitzer. BITZER Software, Version 6.7.0.1849. 2017. https://www.bitzer.de/de/de/software/ [Accessed: November 2, 2017]

[16] Llopis R, Cabello R, Sánchez D, Torrella E. Energy improvements of CO_2 transcritical refrigeration cycles using dedicated mechanical subcooling. International Journal of Refrigeration. 2015;**55**:129-141. DOI: 10.1016/j.ijrefrig.2015.03.016